# Natural Computing Series

Series Editors: G. Rozenberg
Th. Bäck A.E. Eiben J.N. Kok H.P. Spaink

Leiden Center for Natural Computing

T0254695

Springer
*Berlin*
*Heidelberg*
*New York*
*Hong Kong*
*London*
*Milan*
*Paris*
*Tokyo*

Andrzej Ehrenfeucht   Tero Harju
Ion Petre   David M. Prescott
Grzegorz Rozenberg

# Computation in Living Cells

## Gene Assembly in Ciliates

With 92 Figures and 2 Tables

 Springer

Andrzej Ehrenfeucht
Department of Computer Science
University of Colorado
Boulder, CO 80309-0347, USA
email: andrzej@cs.colorado.edu

Tero Harju
Department of Mathematics
University of Turku
FIN-20014 Turku, Finland
email: harju@utu.fi

Ion Petre
Department of Computer Science
Åbo Akademi University
FIN-20520 Turku, Finland
email: ipetre@abo.fi

David M. Prescott
Department of Molecular, Cellular and
Developmental Biology
University of Colorado
Boulder, CO 80309-0347, USA
email: prescotd@colorado.edu

Grzegorz Rozenberg
Leiden Institute for Advanced Computer Science
Leiden University
Niels Bohrweg 1
2333 CA Leiden, The Netherlands
e-mail: rozenber@liacs.nl

*Series Editors*

G. Rozenberg (Managing Editor)
rozenber@liacs.nl
Th. Bäck, J. N. Kok, H. P. Spaink

Leiden Center for Natural Computing, Leiden University
Niels Bohrweg 1, 2333 CA Leiden, The Netherlands

A. E. Eiben
Vrije Universiteit Amsterdam

Cataloging-in-Publication Data applied for

A catalog record for this book is available from the Library of Congress.

Bibliographic information published by Die Deutsche Bibliothek
Die Deutsche Bibliothek lists this publication in the Deutsche Nationalbibliografie; detailed bibliographic
data is available in the Internet at http://dnb.ddb.de.

ACM Computing Classification (1998): F.1, G.2.3, J.3

ISBN 978-3-642-07401-1

Springer-Verlag Berlin Heidelberg New York,
a member of BertelsmannSpringer Science+Business Media GmbH
http://www.springer.de

© Springer-Verlag Berlin Heidelberg 2010
Printed in Germany

The use of general descriptive names, registered names, trademarks, etc. in this publication
does not imply, even in the absence of a specific statement, that such names are exempt from
the relevant protective laws and regulations and therefore free for general use.

Cover design: KünkelLopka, Heidelberg

Printed on acid-free paper    45/3142XO – 5 4 3 2 1 0

# Preface

Natural Computing is a research area concerned with computing taking place in nature and with human-designed computing inspired by nature. It is a fast growing, genuinely interdisciplinary field involving, among others, biology and computer science.

The contribution of Natural Computing to computer science is quite significant, and it comes in the period when computer science is undergoing an important transformation that combines knowledge about human-design computing (as going on in computer science) with knowledge about computing observed in nature. Several areas of natural computing, such as evolutionary algorithms (see, e.g., Ghosh and Tsutsui [23]), neural networks (see, e.g., Haykin [27]), quantum computing (see, e.g., Hirvensalo [29]), and DNA computing (see, e.g., Păun et al. [41]; Jonoska and Seeman [30]), are flourishing in computer science. Characteristic for these areas is the use of paradigms underlying natural systems. Thus, e.g., evolutionary algorithms use the concepts of mutation, recombination, and natural selection from the theory of evolution, while neural networks are inspired by the highly interconnected neural structures in the brain and nervous system.

DNA computing is based on paradigms from molecular biology; researchers in DNA computing study the use of DNA (and other) molecules for the purposes of computing. Research in DNA computing can be roughly divided into two (not disjoint) streams: DNA computing in vitro and DNA computing in vivo. The former is concerned with the theoretical foundations and experimental work on building DNA-based computers in test tubes. The latter is concerned with constructing computational components in living cells (such as simple switching circuits, see [57]), and with studying computational processes taking place in living cells. In recent years, some of the life processes going on in ciliates have attracted the attention of researchers in the DNA computing community.

Ciliates (ciliated protozoa) are single-celled eukaryotic organisms (see, e.g., [47]). It is an ancient group of organisms that originated around two billion years ago, and it is a very diverse group – some 8,000 different species are currently known. Two characteristics unify ciliates as a single group: the possession of hairlike cilia used for motility and food capture, and the presence of two kinds of functionally different nuclei in the same cell – a micronucleus and a macronucleus.

The macronucleus is the "household nucleus" that provides RNA transcripts for producing proteins, while the micronucleus is a dormant nucleus, where no production of RNA transcripts is attempted at all. The micronucleus is activated only in the process of sexual reproduction, where at some stage (the genome of) the micronucleus gets transformed into (the genome of) the macronucleus in the process called gene assembly – it is the most involved DNA processing known in living organisms. Gene assembly is so involved because the form of the micronuclear genome is drastically different from the form of the macronuclear genome (see, e.g., [45], [46], [48]).

The computational nature of gene assembly in ciliates was brought to the attention of the DNA computing community in a series of papers by Kari and Landweber (see, e.g., [37], [38]), where the authors note that the process of assembling a macronuclear gene from its micronuclear form resembles the structure of the solution of the so-called directed Hamiltonian path problem proposed by Adleman in his seminal paper [1] that invigorated DNA computing research. (See also [55] for an even earlier hint on the computational nature of gene assembly.) Since then research on the computational nature of gene assembly in ciliates has developed rapidly, and it has involved both biologists and computer scientists. One line of this research has followed the original view of Kari and Landweber, and it has focused on the computational power (in the sense of computability theory) of their intermolecular model. The other line of this research, carried out by the authors of this book and based on an intramolecular model, has focused on the gene assembly itself, including topics such as the possible forms of the genes generated during gene assembly and possible strategies for the gene assembly. This book centers on the phenomena of gene assembly.

DNA computing represents one side of the cooperation/interaction between computer scientists and biologists: molecular biology assisting computer scientists to achieve the really bold goal of providing an alternative to or a complement for silicon-based computers. On the other side of this cooperation/interaction, in bioinformatics (see, e.g., Lesk [39]) and in computational molecular biology (see, e.g., Pevzner [43]), computer scientists and mathematicians assist biologists in understanding the structure and function of biomolecules, such as DNA and proteins, in living cells. The research presented in this book lies at the intersection of all three areas: DNA computing, bioinformatics, and computational biology. But most naturally it belongs to natural computing because it is deeply concerned with the computational nature of complex biological phenomena.

This book is organized as follows.

Part I of the book gives the biological background, and it consists of three chapters. Chapter 1 provides an overview of the structures common to cells and the molecular principles on which cells operate. Chapter 2 describes the features of ciliates that make them uniquely useful for the study of natural computing. In Chapter 3 we postulate three molecular operations, *ld*, *hi*, and *dlad*, that accomplish gene assembly in ciliates.

Part II introduces formal models for studying gene assembly. Chapter 4 describes the process of model forming that leads to the formulation of three models: MDS descriptors, legal strings, and overlap graphs. In particular, it explains how abstracting from more details of gene structure leads to these three models (in this increasing level of abstraction). This chapter is an informal introduction to the formal framework of this book — it lays a foundation for biologists to acquire intuitive insights and understanding about the more formal chapters of Part II. Chapter 5 introduces basic mathematical notions and notations needed in this book. Formalization of gene assembly on the MDS descriptors level is presented in Chaps. 6 and 7, on the level of legal strings in Chaps. 8 and 9, and on the level of overlap graphs in Chaps. 10 and 11.

Part III gives three examples of research topics concerned with gene assembly. In Chapter 12 we consider properties of gene assembly that are independent of the choice of gene assembly strategy. Since at present we do not know which strategies are actually used by ciliates, studying properties that are common to all strategies is, of course, important. In Chap. 13 we analyze the influence of molecular operations on the form of the genes that they assemble. In particular, we give formal characterizations of the forms of genes that can be assembled by each subset of the set of the three molecular operations *ld*, *hi*, and *dlad*. In Chap. 14 we use graph theory for formulating yet another point of view on gene assembly. We view it here as a process of dynamically changing decomposition of a graph representing a gene. One can view this chapter as a structural graph-theoretic formulation of the novel paradigm "computing by folding and recombination" that underlies a big part of research on computational aspects of gene assembly.

Finally, Part IV is an epilogue for this book. Chapter 15 demonstrates how to formulate an intermolecular model of gene assembly using the "pointer approach" of this book. In this way we formulate one possible bridge to the original intermolecular model of Kari and Landweber. Chapter 16 provides a perspective on the research presented in this book, and in particular it outlines a number of possible future lines of research.

**Acknowledgements**

The research of T. Harju, I. Petre, and G. Rozenberg was supported by European Union project MolCoNet, IST-2001-32008. T. Harju gratefully acknowledges support by the Academy of Finland, project 39802. D. M. Prescott and G. Rozenberg gratefully acknowledge support under NSF grant 0121422. We are also grateful to Springer-Verlag, in particular Mrs. Ingeborg Mayer, for cooperation, excellent in every respect.

August 2003                                             A. Ehrenfeucht
Boulder, Turku, Leiden                                  T. Harju
                                                        I. Petre
                                                        D.M. Prescott
                                                        G. Rozenberg

# Contents

# Notation

## Sets

| | | |
|---|---|---|
| $\mathrm{card}(X)$ | Number of elements in a set | 57 |
| $[k, n]$ | Interval $\{k, k+1, \ldots, n\}$ | 57 |
| $X \cup Y, X \cap Y$ | Union and intersection of sets | 57 |
| $X \setminus Y$ | Set difference | 57 |
| $\mathbb{N}$ | Set of positive integers | 57 |
| $\emptyset$ | Empty set | 58 |
| $\delta_{(p)}$ | Interval of a pointer in $\delta$ | 71 |
| $O_u(a), O_u^+(a), O_u^-(a)$ | Overlapping pointers of a legal string | 84 |
| Ld, Hi, Dlad | Sets of the operations ld, hi, dlad | 76 |
| Snr, Spr, Sdr | Sets of the operations snr, spr, sdr | 92 |
| Gnr, Gpr, Gdr | Sets of the operations gnr, gpr, gdr | 110 |

## Strings

| | | |
|---|---|---|
| $\Sigma^*$ | Set of strings over $\Sigma$ | 58 |
| $\overline{\Sigma}$ | A signed copy of $\Sigma$ | 60 |
| $\Sigma^{\circledast}$ | Set of signed strings over $\Sigma$ | 60 |
| $\mathrm{dom}(u)$ | Domain of the string $u$ | 60 |
| $\|v\|$ | Removes the bars from $v$ | 61 |
| $[v]$ | Circular string of $v$ | 61 |
| $u_{(a)}$ | $a$-interval of a legal string $u$ | 83 |
| $\Lambda$ | Empty string | 58 |
| $\overline{u}$ | Inversion of the string $u$ | 60 |
| $u^R, u^C$ | Reversal and complement of $u$ | 60 |
| $\Theta_\kappa$ | Alphabet of MDSs | 68 |
| $\Omega$ | Alphabet of IESs | 122 |
| $\Delta_\kappa$ | Alphabet $\{2, 3, \ldots, \kappa\}$ | 69 |
| $\mathcal{M}$ | Markers $\{b, e, \overline{b}, \overline{e}\}$ | 69 |
| $\Pi_\kappa$ | Alphabet of pointers $2, \overline{2}, \ldots, \kappa, \overline{\kappa}$ | 69 |
| $\Gamma_\kappa$ | Alphabet of MDS descriptors | 70 |
| $w_\delta$ | String associated with an MDS descriptor | 91 |

## Functions

## Graphs

This book is dedicated to Pat, Eija, Luigia, Gayle, and Maja

# Part I

## Biological Background

# 1

# An Overview of the Cell

All organisms are composed of cells. Some organisms, like an amoeba or a bacterium, consist of a single cell (unicellular organisms). Plants and animals consist of a few hundred to many trillions of cells (multicellular organisms). The structures common to cells and the molecular principles on which cells operate are described in this chapter. The structures and functions of cells are prescribed by genes contained in the DNA molecules of chromosomes. The genetic information in DNA is transcribed into RNA molecules, which in turn are translated into protein molecules. Protein molecules are the principal molecules that prescribe cell structures and functions. Cells duplicate by growing and replicating the DNA in their chromosomes. They distribute the duplicated copies of chromosomes (DNA) to daughter cells, which form by pinching of a cell into two cells.

## 1.1 Cells

The foundations of cell biology were established in the nineteenth century with two major precepts: all organisms, from the simplest to the most complex, are composed of cells, and all cells arise by the division of preexisting cells. Many millions of kinds of organisms, like bacteria and protozoa, are composed of single cells; hence the term single-celled, or unicellular organisms. Many other species are multicellular, in some cases consisting of a few hundred cells; others are formed of trillions ($10^{12}$) of cells. For example, an adult human contains more than $10^{14}$ cells, all derived by a huge number of cell divisions from one initial cell, the fertilized ovum. Cells in the human body are constantly dying and are replaced by more than $25 \times 10^6$ cell divisions per second (producing more than $2 \times 10^{12}$ cells every 24 hours). The many different structural/functional types of cells in a human body are all genetically identical, i.e., contain precisely the same DNA sequences. The differences are created by activating various combinations of genes among the 30,000 or so genes that are common to every cell in the body.

Evolution has given rise to the millions of genetically different organisms that populate the Earth. The number of these contemporary species is not known, but identifications of organisms by analysis of their DNA makes it clear that there are far more species than previously estimated, probably many hundreds of millions. The populations of species is in flux over evolutionary time, with new species arising by genetic (DNA) changes and older species becoming extinct. Our perception of the Earth's organismic population is skewed by the presence of species of multicellular organisms, which strongly tend to dominate our field of view. Vastly more species, occupying a great myriad of microenvironments, make up the unicellular organisms, and evolution is probably a more robust process among these creatures.

In spite of the great genetic diversity that underlies the different species, all organisms, unicellular and multicellular, operate on the basis of the same set of molecular principles. For example, genes in all cells consist of DNA sequences, which when activated are transcribed in RNA molecules. RNA molecules are then translated into protein molecules. Differences among cell species rest on structural/functional differences in proteins prescribed by the different combinations of genes in the different species. As discussed later in this chapter, the number of possible genes, and hence the number of possible different organisms, is enormous.

All cells belong to one of two classes: (1) bacteria, which are all unicellular, do not have a membranous envelope around their chromosome(s) to separate them from the rest of the cell, the cytoplasm; (2) eukaryotes (eu=true, karyote=nucleus), which include both unicellular and multicellular organisms, have the chromosomes aggregated into a body surrounded by a membranous envelope built of molecules called lipids – thus this envelope forms a structurally well-defined cell nucleus.

The remainder of this chapter concerns those elements and principles that are common to the structure and function of cells. Note that structures in cells are composed of combinations of proteins, lipids, and nucleic acids. A protein is a macromolecule (a polymer composed of thousands of atoms) consisting typically of several hundreds of small molecules (monomers composed of tens of atoms) called amino acids. There are 20 different amino acids in proteins. There are two kinds of nucleic acids: DNA (deoxyribonucleic acid) and RNA (ribonucleic acid). DNA is built from monomers called nucleotides, which are of four kinds, designated as A, T, G, and C, standing for *adenine, thymine, guanine*, and *cytosine*, respectively. RNA is also built from nucleotides, again of four kinds, A, U, G, and C, where U stands for *uracil*. Note that A, G, and C are found in both, but DNA contains T's and RNA contains U's instead of T's. By interacting, these large molecules form stable, specific aggregates, which then accomplish one or another function.

Figure 1.1a contains a diagram of the generalized structure of a bacterial cell and of a eukaryotic cell. Bacterial cells (Fig. 1.1a) are structurally rather simple, consisting of a cell membrane, ribosomes, a nucleoid or chromatin body containing the chromosome(s), and the cytoplasmic fluid that fills the

Cell wall   Plasma membrane

(a)

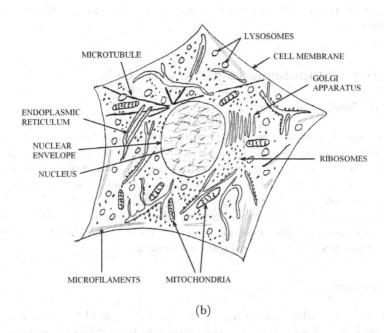

(b)

**Fig. 1.1. a** A section through a bacterial cell (*E. coli*) observed with an electron microscope. The cell is enclosed by a wall (*thick arrow*). The plasma membrane (*thin arrow*) is just inside the cell wall. The granular appearance is due to ribosomes (*R*). The lighter regions are portions of the nucleoid (*N*) which is separating into two parts. The cell is about to divide, as indicated by the slight indentation of the cell wall in the middle of the cell. Courtesy of N. Nanninga [40]. **b** Diagram of a generalized eukaryotic cell. In addition to the structures shown, all of which are present in almost all eukaryotic cell types, some cell types contain other structures, such as chloroplasts (plant and algal cells), fat vacuoles (fat cells), cilia (protozoans and bronchial lung cells), a flagellum (sperm cells), food vacuoles, and so on.

cell. Some bacterial species possess flagella, which give the cell swimming ability. Bacterial cells are generally no more than a few microns in their longest dimension. Eukaryotic cells (Fig. 1.1b) are structurally more complex and larger — ranging from a few microns to hundreds of microns in their longest dimension. They consist of a cell membrane, a well-defined nucleus containing the chromosomes and delineated by a membranous envelope, ribosomes, cytoplasmic fluid, and a series of membranous structures in the cytoplasm, principally the endoplasmic reticulum, the Golgi apparatus, and lysosomes. They also contain two kinds of fiber: microtubules and microfilaments. In addition, some eukaryotic cells are endowed with other structures such as a tough cell wall (e.g., fungi and plant cells), cilia or flagella for motility, lipid droplets in the cytoplasm, etc. We consider here only those structures common to all eukaryotic cells.

## 1.2 Major Components of Eukaryotic Cells

### The Cell Membrane

A cell is enclosed by a thin membrane composed of lipid and protein molecules. This membrane keeps the cell intact, acting as a barrier that retains ions, molecules, and structures within the cell and excludes substances in their microenvironment. The cell membrane also acts to regulate traffic of certain molecules and ions in and out of the cell. For example, some proteins within the cell membrane work as pumps to bring glucose and other nutrients into a cell and to carry inorganic ions out, generally maintaining the physiological health and balance in the intracellular environment.

### The Ribosome

A ribosome is a roughly spherical particle that is too small to see in a light microscope but is clearly visible in the cytoplasm by electron microscopy. A small cell like the bacterium *E. coli* contains more than 20,000 ribosomes. Larger eukaryotic cells may each possess 100,000 or more ribosomes. An individual ribosome consists of about 50 different protein molecules and several RNA molecules (called rRNA molecules). The rRNA molecules provide a scaffold for assembly of the protein molecules arranged in a specific pattern. The entire aggregate, guided by the nucleotide sequences in messenger RNA (mRNA) molecules adhering to the ribosome, assembles amino acids into all the different proteins in a cell, a cytoplasmic process called translation (of mRNA). The many different mRNA molecules are synthesized on DNA templates (genes) in the nucleus (a process called transcription) and then migrate to the cytoplasm to attach to ribosomes.

## The Mitochondrion

Mitochondria are membranous organelles present in the cytoplasm of eukaryotic cells. The membranes of mitochondria are composed of lipid molecules and a specific spectrum of protein molecules, which, in concert with proteins dissolved in the fluid phase of the mitochondrion, carry out the terminal part of the breakdown of fuel molecules like glucose to produce adenosine triphosphate (ATP). ATP is the energy molecule used to drive all of the biosynthetic and other activities (e.g., cell motility) of the cell.

## The Endoplasmic Reticulum

The endoplasmic reticulum (ER) is an extensive system of membranes in the cytoplasm of eukaryotic cells. The lipid and protein molecules that make up the ER membranes perform various metabolic functions, including the transfer of newly synthesized proteins from ribosomes to the Golgi apparatus (see the next paragraph).

## The Golgi Apparatus

The Golgi apparatus consists of a stack of flattened membranous sacs in the cytoplasm of eukaryotic cells. It receives various proteins from the ER, modifies certain of these proteins, e.g., by appending particular sugar molecules to them, and packages the proteins into vesicles for delivery to various other intracellular destinations or for excretion from the cell.

## Microtubules and Microfilaments

Microtubules are long tubular structures in the cytoplasm, visible only by electron microscopy, composed of thousands of molecules of a single kind of protein. Microfilaments are long cytoplasmic rodlike filaments, visible only by electron microscopy, composed of thousands of molecules of another kind of protein. Both microtubules and microfilaments often occur in bundles, and both are major elements of the skeletal framework of the eukaryotic cell. They both have roles in cell movement, e.g., the flowing of cytoplasm as occurs in amoeboidlike movement, the movement of chromosomes during cell division, and the contraction that cleaves a cell in two in cell division.

## The Lysosomes

The cytoplasm of eukaryotic cells possesses numerous membrane-bound vesicles called lysosomes. Lysosomes contain a variety of digestive enzymes (proteins) that break down large molecules that may have been ingested by a cell. For example, many protozoans (like an amoeba or a ciliate) make a living by ingesting other microorganisms, both bacteria and eukaryotes, a process called phagocytosis. These ingested food organisms are enveloped in an infolding of the cell membrane to form a vesicle, called a food vacuole. Lysosomes fuse with the food vacuoles, and the lysosomal enzymes break down the proteins, nucleic acids, and lipids of the food organism into smaller molecules, which fuel the metabolism of the cell. Phagocytosis is a major mechanism by which white blood cells ingest and destroy (by means of lysosomal enzymes) bacteria that may have infected a multicellular animal, e.g., a human.

### The Cytoplasmic Solution

The structures in the cytoplasm are immersed in a cytoplasmic solution or broth containing a broad spectrum of chemical components, protein molecules, inorganic ions ($Ca^{++}$, $Na^+$, $K^+$, $Cl^-$, $PO_4^-$, etc.), sugar molecules, amino acids, nucleotides, and many other small organic molecules. These dissolved components support many functions, e.g., amino acids and nucleotides serve as precursor units in the synthesis of proteins and nucleic acids (DNA and RNA), respectively. An important role of the cytoplasmic solution is the coordination and integration of the activities of the structures in the cell, including the nucleus.

These are the major components of eukaryotic cells, although there are numerous other structures and molecular elements that perform specialized functions in a variety of eukaryotic cell types, e.g., cilia, which are hairlike projections in the cell surface that beat in synchronous rhythm to propel a cell through its aqueous environment. Cilia (singular=cilium) are a defining characteristic of a large group of unicellular protozoan species known as ciliates. Ciliates are described in more detail Chap. 2. Much of this book is based on the genetic processes in ciliates.

## 1.3 Chromosome Structure

Chromosomes are extremely long threadlike structures that contain the DNA molecules of the cell. They form a tangled meshwork in the cell nucleus, making it impossible to discern individual chromosomes even by electron microscopy. However, transiently during cell division chromosomes condense into short thick rods that can be seen even in a light microscope. In this condensed form the chromosomes are distributed to the two daughter cells during cell division, after which they return to their indiscernible, decondensed state. Chromosomes contain the genes in a cell, and in that sense represent the master blueprint for all cell structures and operations. The number of chromosomes, each with its different collection of genes, is characteristic of a cell species. A fruit fly cell has 4 chromosomes, a human cell has 23 different chromosomes, and one species of amoeba has several hundred.

The central element of a chromosome is a single enormously long DNA molecule. Recall that a DNA molecule is a polymer consisting of monomers, called nucleotides. These building blocks are joined together in a long chain (Fig. 1.2a). The nucleotides are of four types, designated A, T, G, and C. Each nucleotide is, in turn, made up of three molecules: a sugar molecule (deoxyribose sugar denoted by S), a phosphate group (P), and a base. The four different nucleotides differ only in their bases, designated A, T, G, and C, giving the nucleotides their names. In a DNA strand the nucleotides are joined in chain fashion through connections between their sugar–phosphate portions (Fig. 1.2b), leaving the bases projecting from the side of the sugar–phosphate backbone.

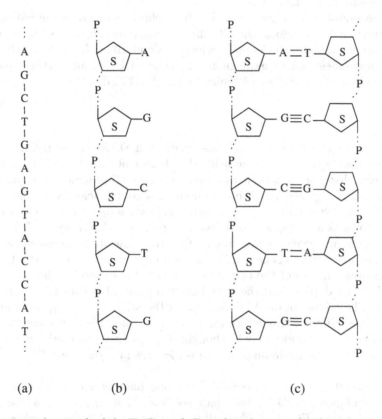

(a)            (b)                    (c)

**Fig. 1.2. a** A strand of A, T, G, and C nucleotides in a DNA molecule. **b** A segment of a DNA strand. Nucleotides in DNA are joined to form very long strands by connections between the sugar (S) and phosphate (P) groups in each nucleotide. The bases (A, T, G, and C) in the nucleotides project to the side of the strand. **c** A segment of a DNA double helix. Two nucleotide strands bind to each other by forming base pairs between A and T and between G and C. The two strands are shown here in a simple parallel arrangement. In a double helix the two chains twist around one another in a helical fashion.

The DNA molecule in a chromosome of a bacterium or eukaryotic cell consists, in fact, of two DNA strands wound around each other in helical fashion, forming what is commonly referred to as a DNA double helix or duplex (Fig. 1.2c). The two strands in a duplex interact with each other through binding of the bases in one strand with the bases in the other strand. These interactions of the bases are highly specific, being dictated by the different chemical structures of the four bases. Thus, wherever an A occurs in one strand, it is bound to a T in the other strand; wherever a G occurs in a chain, it is paired with a C in the opposite strand. Each of these A to T (or T to A) and G to C (or C to G) couplings of nucleotides through their bases is called a base pair, and the two DNA strands in a double helix are said to be complementary to one another.

DNA double helices vary in length from chromosome to chromosome; the longest DNA helix among the 23 different chromosomes in a human cell is made up of $\sim 3 \times 10^8$ nucleotides, which is about 10 cm long. Altogether the 23 chromosomes contain more than one meter of DNA; all of this is packed into a spherical nucleus 8 to 10 microns ($10^{-6}$ m) in diameter.

**Genes**

A gene is a specific sequence of base pairs in a DNA double helix. A gene specifies the sequence of amino acids that is assembled into a chain to make a protein. How this is done is discussed in this section under gene function. The length of a gene depends on the number of amino acids in the particular protein it specifies, and the gene lengths vary over a broad range. On average a gene consists of a segment of 1,000 base pairs in a DNA double helix, and there are many genes in the single DNA molecule in a chromosome. The beginning and end of a gene within a DNA double helix are marked by triplets of nucleotides in one of the two DNA strands; this strand is called the sense strand of the duplex, and the complementary strand is called the antisense strand. The trinucleotide ATG designates the start of every gene, and any one of three different trinucleotides, TGA, TAA, and TAG, specifies the end. These four trinucleotides may be thought of as punctuation marks in a double helix, marking the beginnings and ends of genes in a very long DNA double helix.

Individual genes are separated from one another along DNA by long stretches of nongene DNA base pair sequence that forms spacers between successive genes (Fig. 1.3). Typically, less than 5% of a DNA molecule encodes genes; more than 95% makes up the spacers. Often, spacer DNA is referred to as junk DNA because it has no known function other than separating successive genes in a chromosome by wide distances. If and why such wide separation of genes is in some way important is not known. Nevertheless, a single DNA double helix in a eukaryotic cell may contain many hundreds of genes; the 23 DNA double helices in a human cell contain a total of about 30,000 genes.

**Fig. 1.3.** The arrangement of genes along a segment of eukaryotic chromosomal DNA. Less than 5% of the DNA molecule encodes genes. Most of the DNA molecule acts as spacers between successive genes.

## Diploidy

Most species of eukaryotic cells are diploid that is, each cell contains two copies of every chromosome. For example a human cell has 23 different chromosomes but two copies of each, for a total of 46. One set of 23 was derived from the female parent (an ovum has 23 chromosomes), and one set of 23 from the male parent (a sperm carries 23 chromosomes). Fertilization of an ovum by a sperm creates a cell (the fertilized ovum) with 46 chromosomes, and the vast majority of cells in the body are diploid (the exceptions are ova and sperm cells). Mating can occur between two unicellular organisms that belong to the same species. The process is essentially the same as in multicellular organisms like the human. Mating of unicellular organisms is explained in Chap. 2 about ciliates.

## Gene Function

A gene exerts its role in a cell through the protein that it encodes, i.e., proteins are the metabolically active molecules in a cell because of their catalytic and regulatory properties. The DNA sequence in a gene prescribes the sequence of amino acids in the protein that the gene encodes by means of two steps: transcription and translation.

In transcription, the DNA in the gene guides in template fashion the synthesis of an RNA molecule, called a messenger, or mRNA molecule. This is accomplished by transient separation of the two strands in the DNA of the gene. The antisense strand serves as a template to assemble an mRNA molecule using base-pairing rules (Fig. 1.4). Wherever a T occurs in the antisense strand of the gene, an A is added to the mRNA molecule. G is similarly matched by C, and C is matched by G. However, an A in the gene designates a U in the mRNA. In effect, the antisense strand of the gene is complementary to the single-stranded mRNA molecule, except that T is substituted by U in mRNA. The T and U nucleotides are chemically very similar. Thus the sequence of nucleotides in the mRNA is identical to the sense strand of the gene, with the exception that the T's in DNA are substituted with U's in the mRNA. All of this happens very quickly, with many nucleotides per second added to the growing mRNA molecule. These additions are catalyzed by an enzyme (a protein) called RNA polymerase.

Completed mRNA molecules (after several modifications) are then transported to the cytoplasm, where they bind to ribosomes for translation. In

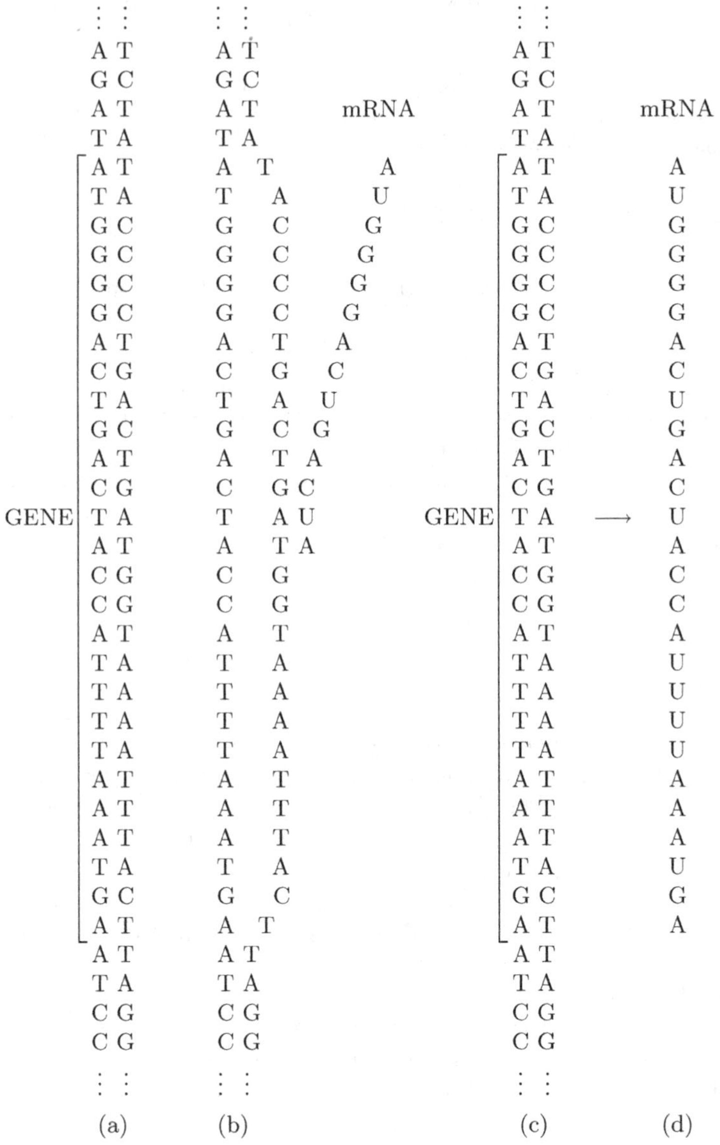

| | (a) | (b) DNA | (b) mRNA | (c) DNA | (d) mRNA |
|---|---|---|---|---|---|
| | ⋮ ⋮ | ⋮ ⋮ | | ⋮ ⋮ | |
| | A T | A Ṫ | | A T | |
| | G C | G C | | G C | |
| | A T | A T | mRNA | A T | mRNA |
| | T A | T A | | T A | |
| | A T | A T | A | A T | A |
| | T A | T A | U | T A | U |
| | G C | G C | G | G C | G |
| | G C | G C | G | G C | G |
| | G C | G C | G | G C | G |
| | G C | G C | G | G C | G |
| | A T | A T | A | A T | A |
| | C G | C G | C | C G | C |
| | T A | T A | U | T A | U |
| | G C | G C | G | G C | G |
| | A T | A T | A | A T | A |
| | C G | C G | C | C G | C |
| GENE | T A | T A | U | T A (GENE) | → U |
| | A T | A T | A | A T | A |
| | C G | C G | | C G | C |
| | C G | C G | | C G | C |
| | A T | A T | | A T | A |
| | T A | T A | | T A | U |
| | T A | T A | | T A | U |
| | T A | T A | | T A | U |
| | T A | T A | | T A | U |
| | A T | A T | | A T | A |
| | A T | A T | | A T | A |
| | A T | A T | | A T | A |
| | T A | T A | | T A | U |
| | G C | G C | | G C | G |
| | A T | A T | | A T | A |
| | A T | A T | | A T | |
| | T A | T A | | T A | |
| | C G | C G | | C G | |
| | C G | C G | | C G | |
| | ⋮ ⋮ | ⋮ ⋮ | | ⋮ ⋮ | |

(a)          (b)                    (c)          (d)

**Fig. 1.4. a–d** Transcription of a gene into an mRNA molecule: **a** A gene in a DNA duplex. The gene begins with ATG and ends in TGA (stop). **b** The two strands of DNA in the gene have separated, and the antisense strand is serving as a template for assembling an mRNA molecule. The mRNA is only partially formed, separating from its template as it is synthesized. Note that T's in DNA are substituted by U's in mRNA. **c** Transcription has finished and the DNA of the gene has returned to its duplex state. **d** The finished mRNA molecule. Note that the TGA (stop) in the gene is not transcribed into the mRNA molecule.

translating mRNA the ribosome begins with the AUG (equivalent to the ATG at the beginning of the sense strand in the DNA of a gene). The ribosome translates the AUG as the amino acid methionine. Thus, all amino acid chains of proteins begin with methionine. The ribosome moves to the next three nucleotides in the mRNA, interprets them as one of the other 19 amino acids, and joins that amino acid to the methionine. The ribosome translates each successive trinucleotide as a particular amino acid, adding to the growing amino acid chain. Finally, at the end of the mRNA the ribosome encounters UGA, UAA, or UAG and translates this as "stop" and releases the finished amino acid chain (protein). As a ribosome leaves the AUG site and progresses along the mRNA, another ribosome binds to AUG at the beginning of the mRNA; thus, multiple ribosomes are translating the mRNA at any one time, producing multiple identical copies of the particular protein product.

**Table 1.1.** The genetic code table

| First position (5′ end) | Second position | | | | Third position (3′ end) |
|---|---|---|---|---|---|
| | U | C | A | G | |
| U | Phe | Ser | Tyr | Cys | U |
| | Phe | Ser | Tyr | Cys | C |
| | Leu | Ser | Stop | Stop | A |
| | Leu | Ser | Stop | Trp | G |
| C | Leu | Pro | His | Arg | U |
| | Leu | Pro | His | Arg | C |
| | Leu | Pro | Gln | Arg | A |
| | Leu | Pro | Gln | Arg | G |
| A | Ile | Thr | Asn | Ser | U |
| | Ile | Thr | Asn | Ser | C |
| | Ile | Thr | Lys | Arg | A |
| | Met | Thr | Lys | Arg | G |
| G | Val | Ala | Asp | Gly | U |
| | Val | Ala | Asp | Gly | C |
| | Val | Ala | Glu | Gly | A |
| | Val | Ala | Glu | Gly | G |

The trinucleotides in the mRNA (or in the sense strand) of the DNA of the gene that prescribe one or another of the 20 amino acids are called codons. Thus, a gene consists of a succession of codons that the cell reads, via mRNA, into a sequence of amino acids. The four bases, A, T, G, and C (or A, U, G, and C in the mRNA), can be combined into 64 different trinucleotide codons. Three of these codons, ATG (AUG in mRNA), TGA (UGA in mRNA), and TAA (UAA in mRNA), are used to define the end of a gene (or the mRNA), and one codon, ATG (AUG in mRNA), defines the amino acid methionine at the beginning of a gene. The remaining 60 triplet codons are unevenly divided among the other 19 amino acids, e.g., the amino acid tryptophan is designated by one codon, isoleucine is designated by three different codons in a gene, and arginine is designated by six different codons; see the genetic code table (Table 1.1), and the table of amino acid names and their three letter abbreviations (Table 1.2). An example of a protein encoded by the sequence of codons of a gene is given in Fig. 1.5.

**Table 1.2.** The 20 amino acids and their three-letter abbreviations

| Name | 3-letter abbreviation | Name | 3-letter abbreviation |
|---|---|---|---|
| Alanine | Ala | Leucine | Leu |
| Arginine | Arg | Lysine | Lys |
| Asparagine | Asn | Methionine | Met |
| Aspartic acid | Asp | Phenylalanine | Phe |
| Cysteine | Cys | Proline | Pro |
| Glutamine | Gln | Serine | Ser |
| Glutamic acid | Glu | Threonine | Thr |
| Glycine | Gly | Tryptophan | Trp |
| Histidine | His | Tyrosine | Tyr |
| Isoleucine | Ile | Valine | Val |

In summary, through gene expression, i.e., the process of transcription and translation, the many hundreds of different proteins needed to form the structures and support the metabolism in a cell are continuously produced.

## 1.4 Chromosomes and Genes

Cells reproduce by cell division. In order to divide, a cell must first increase in size so that two daughter cells of adequate size can be produced. A major

| amino acids | Met | Ala | Gly | Arg | Gly | Lys | Val | Gly | Lys | Gly | Tyr |
|---|---|---|---|---|---|---|---|---|---|---|---|

amino acids      *Met Ala Gly Arg Gly Lys Val Gly Lys Gly Tyr*
sense strand      ATG GCC GGA AGA GGT AAA GTT GGA AAA GGA TAC →
antisense strand   TAC CGG CCT TCT CCA TTT CAA CCT TTT CCT ATG →

amino acids      *Gly Lys Val Gly Ala Lys Arg His Thr Lys Lys*
sense strand      GGA AAG GTT GGT GCC AAG AGA CAC ACC AAG AAG →
antisense strand   CCT TTC CAA CCA CGG TTC TCT GTG TGG TTC TTC →

amino acids      *Ser Leu Lys Glu Thr Ile Met Gly Ile Thr Lys*
sense strand      TCA CTC AAG GAG ACT ATC ATG GGA ATC ACC AAG →
antisense strand   AGT GAG TTC CTC TGA TAG TAC CCT TAG TGG TTC →

amino acids      *Pro Ala Ile Arg Arg Leu Ala Arg Gly Gly Val*
sense strand      CCA GCA ATC AGA AGA CTC GCC AGA GGT GGT GTC →
antisense strand   GGT CGT TAG TCT TCT GAG CGG TCT CCA CCA CAG →

amino acids      *Lys Arg Ile Ser Ser Leu Ile Tyr Glu Glu Thr*
sense strand      AAG AGA ATC TCA TCC CTC ATC TAT GAG GAG ACC →
antisense strand   TTC TCT TAG CGT AGG GAG TAG ATA CTC CTC TGG →

amino acids      *Arg Asp Val Leu Arg Ser Phe Leu Glu Asn Val*
sense strand      AGA AAC GTC CTC AGA TCA TTC CTC GAG AAC GTT →
antisense strand   TCT TTG CAG GAG TCT AGT AAG GAG CTC TTG CAA →

amino acids      *Ile Arg Asp Ser Val Thr Tyr Thr Glu His Ala*
sense strand      ATC AGA GAT TCA GTC ACC TAC ACT GAA CAC GCC →
antisense strand   TAG TCT CTA AGT CAG TGG ATG TGA CTT GTG CGG →

amino acids      *Lys Arg Lys Thr Val Thr Ala Leu Asp Val Val*
sense strand      AAG AGA AAG ACC GTC ACC GCT CTC GAC GTC GTC →
antisense strand   TTC TCT TTC TGG CAG TGG CGA GAG CTG CAG CAG →

amino acids      *Tyr Ala Leu Lys Arg Gln Gly Arg Thr Leu Tyr*
sense strand      TAC GCT CTT AAG AGA CAA GGA AGA ACC CTC TAC →
antisense strand   ATG CGA GAA TTC TCT GTT CCT TCT TGG GAG ATG →

amino acids      *Gly Phe Gly Gly Stop*
sense strand      GGA TTC GGT GGA TGA
antisense strand   CCT AAG CCA CCT ACT

**Fig. 1.5.** The gene encoding a small protein called histone H4. The sense strand containing the code for the protein is in the *middle*, and the complementary strand is immediately *below* it. The sense strand begins with the three nucleotides ATG, which specifies methionine, and ends with TGA, which means stop, or period. The amino acid chain specified by successive groups of three nucleotides (trinucleotide codons) is shown *above* the DNA double helix.

part of this growth is the synthesis of protein molecules as just described in Sect 1.3. Another essential preparation for cell division is the duplication of the chromosomes so that each daughter cell can receive a full diploid set of chromosomes. In other words, each DNA double helix in every chromosome must first be duplicated if a cell is to divide successfully.

The duplication of DNA molecules is referred to as DNA replication. DNA replication is similar to transcription of DNA in the sense that the DNA serves as a template for making new nucleotide strands, in this case new DNA strands. As in transcription, the two strands in a DNA double helix separate from one another, and the resulting two single strands each serve as a template to guide the polymerization of nucleotides into a complementary strand (Fig. 1.6). The two new complementary strands formed along each of the two template strands remain bound to their template strands, forming in this way two daughter double helices in which one strand in each double helix comes from the original double helix and one strand has been newly synthesized. Of course, the two daughter double helices are identical to each other in nucleotide sequences, and therefore contain precisely the same genes, and are identical to the original parental double helix.

The actual mechanism of DNA replication is a little more complex because the two strands in a double helix have polarities in their sugar–phosphate backbones (Fig. 1.7). In the backbone a phosphate group links the number 3 carbon atom in one sugar (a deoxyribose sugar with five carbon atoms) to the number 5 carbon in the next sugar molecule. One can think of the backbone as sugar molecules linked by phosphate groups through their number 3 and number 5 carbon atoms:

$$C_5\,S\,C_3 - P - C_5\,S\,C_3 - P - C_5\,S\,C_3 - P - C_5\,S\,C_3 - P - \ldots - C_5\,S\,C_3 \ .$$

Thus, a DNA strand has a $C_5$ at one end and a $C_3$ at the other end; these are called the 5' and 3' ends (see Fig. 1.7). In a double helix the two DNA strands have opposite polarities, i.e., at the end of a double helix one strand ends in 5' and the other in 3', as clearly shown in Fig. 1.7. In DNA replication new strands grow continuously by adding nucleotides to a $C_3$ of a sugar. Adding nucleotides to the 5' end is more involved.

Typically, DNA replication in a eukaryotic cell takes several hours. During this time the cell is also growing, increasing all its cytoplasmic components. With the completion of DNA replication and cell growth, the cell undergoes division, producing two genetically identical daughter cells, each of which grows, replicates its DNA, and ultimately divides. Cell division consists of two coordinated events: division of the nucleus, called mitosis, and division of the cytoplasm, called cytokinesis.

### Mitosis and Cytokinesis

Mitosis begins with the condensation of all the replicated chromosomes into short, thick rods that, in most species, are clearly visible in the light mi-

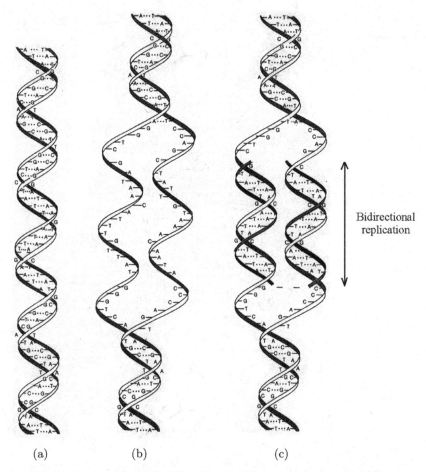

(a)                    (b)                    (c)

**Fig. 1.6. a–c.** Replication of the DNA double helix: **a** A segment of a DNA double helix. **b** The DNA double helix has opened into separated strands in a localized region. **c** The separated strands serve as templates for making new complementary strands.

croscope. The replicated state of the chromosomes is clearly evident; each chromosome consists of two identical rods held together side-by-side along their lengths (Fig. 1.8). The condensed replicated chromosomes line up in an equatorial plate of the cell. Next the two rods in each replicated chromosome separate from one another, forming two identical sets of rods, or daughter chromosomes. One set is drawn to one side of the cell, and the other set is drawn to the other side. These daughter chromosome movements are accomplished by a motility mechanism based on microtubules. Simultaneously, a ring of constriction forms in the cell surface at the equatorial plate, formerly occupied by the chromosomes, and contracts, cleaving the cell into two daugh-

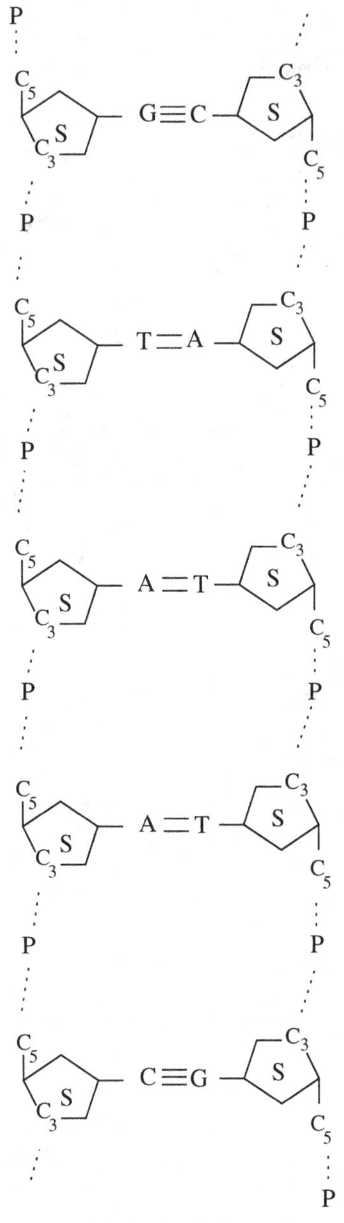

**Fig. 1.7.** Opposite polarities of the two strands in a DNA duplex. Note that at the *upper end* the left strand ends with P attached to the $C_5$ of a sugar (S), and the right strand ends with $C_3$. These are called 5′ and 3′ ends, respectively. At the *lower end* of the duplex the left strand end is 3′ and the right strand end is 5′.

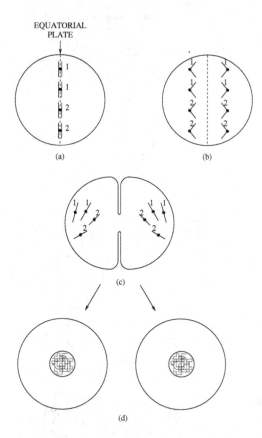

EQUATORIAL
PLATE

(a)

(b)

(c)

(d)

**Fig. 1.8. a–d.** Mitosis: **a** A diploid set of chromosomes. There are two different chromosomes and two copies of each. Each chromosome has been duplicated (two rods in each, aligned next to one another and held together in the middle). **b** The two rods (daughter chromosomes) in each duplicated chromosome have separated and are being drawn to opposite sides of the equatorial plate. **c** A diploid set of chromosomes is present in each incipient cell, and a cleavage furrow through the equatorial plate is about to complete cytokinesis. **d** The chromosomes in the two daughter cells have decondensed into fine fibrous networks, forming a nucleus in each daughter cell.

ters, each with a set of daughter chromosomes (Fig. 1.8c). This cleavage of the cytoplasm is accomplished by a contractile motility mechanism based on microfilaments. As cytokinesis nears completion, the set of chromosomes in each incipient daughter cell quickly decondenses and forms a nucleus, completing the process of cell division (Fig. 1.8d).

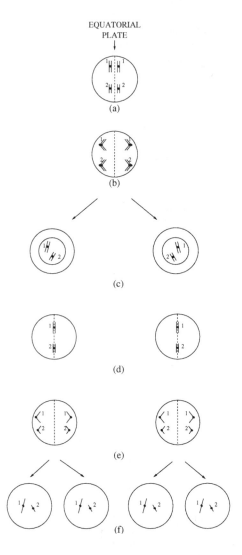

EQUATORIAL
PLATE

**Fig. 1.9. a–f.** Meiosis: **a** The two copies of each chromosome in the diploid set of chromosomes have lined up next to each other. There are two different chromosomes in this diploid set. Each chromosome has been duplicated, i.e., consists of two rods. **b** The chromosomes separate. One duplicated copy of each chromosome is drawn to one side of the equatorial plate and the other is drawn to the opposite side. **c** Cytokinesis has created two haploid daughter cells, i.e., each contains one of each of the two different chromosomes, although the chromosomes are in the duplicated state, i.e., two rods in each. **d** The two different duplicated chromosomes in the two haploid cells in (c) line up in an equatorial plate. **e** The two rods in each of the duplicated chromosomes separate and are drawn to opposite sides in each of the two haploid cells. **f** Cytokinesis of the cells in (e) yields four haploid cells with two different unduplicated chromosomes.

**Formation of Germ Nuclei — Meiosis**

In a multicellular organism like a human, germ nuclei are formed in ova and sperm cells as a prerequisite for cell mating, i.e., fusion of a sperm cell with an ovum. A critical part of the formation of ova and sperm is the reduction from diploidy (two copies of each chromosome) to haploidy (one copy of each chromosome). Thus, when a sperm fuses with an ovum, the two haploid nuclei fuse to form a diploid nucleus and development begins. Reduction from diploidy to haploidy in germ cells (ova and sperm cells) occurs by a modified form of mitosis, called meiosis, in which the two copies of a duplicated chromosome align next to each other in an equatorial plate (Fig. 1.9a) instead of lining up separately, as in mitosis (Fig. 1.8a). The two copies of each chromosome then separate from each other and are drawn to opposite sides of the cell (Fig. 1.9b). Thus, two sets of chromosomes are formed, but unlike in mitosis, every chromosome is still in its double-rod, duplicated form (Fig. 1.9c). Also, each set of duplicated chromosomes now contains one copy of each chromosome (is haploid). The cell completes division, and then each of the two daughter cells immediately divides again, this time in typical mitotic fashion. That is, the haploid set of duplicated chromosomes in each cell lines up in an equatorial plate (Fig. 1.9d), and the two rods in each chromosome separate and migrate to opposite sides of the cell, forming two haploid unduplicated sets of chromosomes (Fig. 1.9e). The two cells divide, producing four haploid daughter cells from one original diploid cell (Fig. 1.9f). Creation of haploid nuclei in preparation for mating in unicellular species is essentially the same. A specific example of this is described for ciliated protozoans in Chap. 2.

## Notes on References

A simple description of cell structure and function, cell reproduction, and DNA structure, function, and replication can be found in Prescott [44]. Also, a good description of these topics can be found in Alberts et al. [2].

# 2

# Ciliates

The features of ciliates that make them uniquely useful for the study of natural computing are described in this chapter. Of particular advantage is the presence in these unicellular organisms of micronuclei and macronuclei. Genes in micronuclear chromosomes are in DNA segments, called MDSs, created by the insertion during evolution of segments of nongenetic DNA, called IESs. In some genes the MDSs have been reordered into scrambled arrangements. After cell mating a micronucleus is converted into a macronucleus, which requires excision of IESs and splicing of MDSs in an orthodox order, processes that are extraordinary cases of natural computing.

## 2.1 Defining Characteristics of Ciliates

Ciliates (ciliated protozoans) form a large group of unicellular organisms ranging in size from a few microns to several hundred microns. Their success can be judged from their worldwide distribution. They inhabit all bodies of fresh and salt water at all latitudes and are found in essentially all soils, even in desert sands during transient periods of moisture. About 8,000 different species are known. It is likely that unknown species far exceed that number. They are an ancient group, originating perhaps as many as two billion ($10^9$) years ago. During their long history, ciliates have evolved enormously into hundreds of different groups. The extent of their diversity can be appreciated by comparing nucleotide sequences of their genes; some ciliate types differ genetically from other types more than fruit flies differ genetically from humans!

Amidst the great genetic diversity of ciliates, two characteristics hold them together as a single group. First, they all possess cilia projecting from their surfaces (Fig. 2.1). The synchronous beating of the cilia propels the organism through its aqueous environment; also, in most kinds of ciliates the beating of cilia drives food organisms (e.g. bacteria, algae, other ciliates) into a pocket in the cell surface, called the oral apparatus, where they are ingested. In addition, ciliates are also characterized by the presence of two kinds of nuclei in the same cell – a micronucleus and a macronucleus – a phenomenon unique to ciliates. The micronucleus is used in mating between ciliates belonging to the same species, and the macronucleus produces all the RNA needed for cell structures and operations.

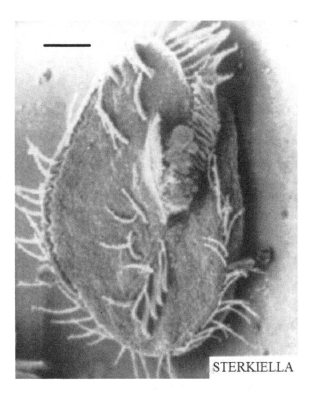

STERKIELLA

**Fig. 2.1.** A scanning electron micrograph of a stichotrich. Note the hairlike projections (cilia) on the cell surface. Bar=10 $\mu$m. Courtesy of Gopal Murti

The molecular genetics of very few ciliate groups have been studied. Extensive research has been done on one group of ciliates known as the stichotrichs (until recently stichotrichs were called hypotrichs). Stichotrichs are of special interest because their DNA has undergone extraordinary modifications during evolution – modifications that require, in turn, unprecedented manipulations of DNA during the day-to-day life of these cells. These recently discovered modifications and manipulations of DNA greatly expand our view of the versatility of DNA changes in biological evolution and in the operation of cells.

## 2.2 Nuclear Dualism

There are hundreds, perhaps thousands, of species in the stichotrich group, but virtually all of the research so far has focused on about 20 different organisms. Most ciliates possess one micronucleus and one macronucleus, but stichotrichs have two or more micronuclei and two or more macronuclei per organism, depending on the species (Fig. 2.2). All the micronuclei are genetically and structurally identical to each other, and all the macronuclei are identical. The significance of this multiplicity of nuclei of each type is not understood.

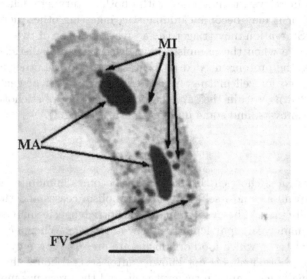

**Fig. 2.2.** Light microscope photograph of a stained stichotrich showing macronuclei, micronuclei, and food vacuoles. This species has two macronuclei (*MA*) and four micronuclei (*MI*). Two food vacuoles (*FV*) are also visible. Bar=10 $\mu$m. Courtesy of Gopal Murti

### The Micronucleus

Micronuclei in stichotrichs and other ciliates contain typical eukaryotic chromosomes, and micronuclei divide by mitosis during cell reproduction. However, micronuclear chromosomes differ dramatically in function from chromosomes in the nuclei of other kinds of eukaryotes. Micronuclear genes are silent, i.e., they are not expressed (not transcribed and translated). Thus, micronuclei lack the major function of synthesizing RNA, and they make no contribution to the growth and reproduction of the cell. The sole function of the micronucleus is in the exchange of genomes during mating between two ciliates. (A genome is all the DNA in a haploid or diploid set of chromosomes.)

All of the products of genes, i.e., RNA transcripts, needed in the cytoplasm to carry out protein synthesis are provided by the genes in the macronuclei; these genes are continuously expressed at a high rate.

## Reaction to Starvation

If nutrients, in the form of food organisms, are available to stichotrichs, they grow and reproduce by cell division. When food organisms become scarce, and a stichotrich begins to starve, it may embark on one of three other fates: it may form an inert cyst, it may mate with another starving stichotrich of the same species, or it may become cannibalistic, ingesting other members of its own species. Starvation may trigger the activation of one of two sets of genes that are inactive when the organism is well fed or cannibalistic. One set of genes encodes the proteins needed to form a cyst, and the other encodes the proteins required for cell mating. What decides which set of genes becomes active is unknown. Within the same laboratory culture of stichotrichs, some organisms form cysts, and some undergo mating (Fig.2.3).

## Cysts

To form a cyst, a stichotrich shrinks in size, becomes immobile, and synthesizes a tough wall around itself. All cell metabolism ceases and the organism becomes totally inert. The cysts in many stichotrich species may survive in a dry state for many years, particularly if they are frozen. When a cyst becomes immersed in water in which food organisms are present (how a cyst senses the presence of food organisms is not known), after several hours the stichotrich emerges through an opening in the cyst wall, and then resumes motility, feeding, and proliferation. Thus, the cyst is a strategy for surviving through long periods when the environment may become unfavorable.

## Cell Mating

Mutations in the DNA of genes caused by chemicals and radiation (such as ultraviolet light, cosmic rays, or radioactive elements), and mistakes in DNA replication occur in all organisms. Such mutations are an essential part of evolution. Most mutations are either deleterious, which may result in the death of the cell, or are neutral, i.e., they have no effect on the function of a gene or on cell survival. On rare occasions a mutation may improve the existing function of a gene or may even give the gene a new function. Such favorable mutations can be spread through a cell population by mating. Mating in unicellular organisms like stichotrichs provides for the exchange of haploid genomes between two cells.

Figure 2.4 shows a stichotrich with two micronuclei and two macronuclei. Mating begins with the sticking together of two organisms (Figs. 2.3 and 2.4b).

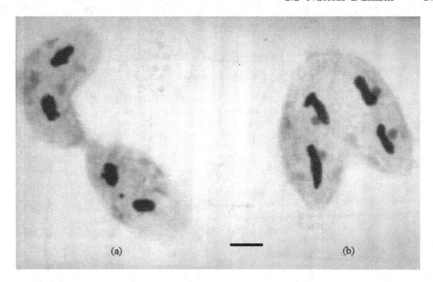

**Fig. 2.3.** Light microscope photograph of a mating pair of stichotrichs. In **a** the two organisms have stuck together and formed a cytoplasmic channel. In **b**, a later stage, the mating pair is more tightly joined. Bar=10 $\mu$m

In the area of sticking the two cell membranes first fuse and a connecting channel is formed (Figs. 2.3a and 2.4b). In coordination with these events the micronuclei in each cell undergo meiosis. Meiosis of each diploid micronucleus yields four haploid micronuclei in the two mating partners. In stichotrichs that possess two diploid micronuclei, eight haploid micronuclei are produced (Fig. 2.4b). Each of the two cells then exchanges one haploid micronucleus with its mating partner through the cytoplasmic channel (Fig. 2.4b). The newly received haploid micronucleus fuses with a resident haploid micronucleus, forming a new diploid micronucleus, and the two cells separate and heal their cell membranes (Fig. 2.4c). The six unused haploid micronuclei and the two macronuclei in each cell begin to degenerate, and at the same time the new diploid micronucleus in each cell divides by mitosis without accompanying division of the cell (Fig. 2.4d). One of the two daughters remains as the new diploid micronucleus. The other daughter micronucleus develops into a new macronucleus during the next three days, during which time the unused haploid micronuclei and the old macronuclei completely disappear (Fig. 2.4e). At the completion of its development the new macronucleus and micronucleus divide (without cell division), yielding the characteristic number of two macronuclei and two micronuclei in the exmated cell and marking the end of the mating process (Fig. 2.4f).

The development of a new macronucleus from a new micronucleus involves many changes in DNA. These changes can begin to be appreciated by examining the structure of micronuclear and macronuclear DNAs.

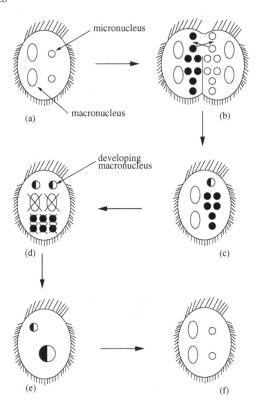

**Fig. 2.4. a–f.** Mating in stichotrichs: **a** A stichotrich with two macronuclei and two micronuclei. **b** Two stichotrichs have joined and formed a cytoplasmic channel. The two diploid micronuclei have each formed four haploid micronuclei. **c** The two cells exchanged haploid micronuclei, and the two organisms have separated. The exchanged haploid micronucleus has fused with a resident haploid micronucleus forming a new diploid micronucleus (half white and half black). **d** The new diploid micronucleus has divided by mitosis. The unused haploid micronuclei (six) and the two macronuclei are degenerating. **e** One of the new daughter micronuclei has developed into a new macronucleus. The old macronucleus and the unused haploid micronuclei have disappeared. **f** Mating has been completed. The micronucleus and macronucleus have divided, yielding the appropriate nuclear numbers (2 and 2).

## 2.3 Micronuclear Versus Macronuclear DNA

The DNA in micronuclear chromosomes, as in chromosomes in general, is very long, and each chromosome contains one DNA molecule. Figure 2.5 shows an electron micrograph of a small section of a micronuclear DNA molecule. Genes are widely spaced along such a molecule, separated by long stretches of nongenetic (spacer) DNA, as in Fig. 1.3 in Chap. 1.

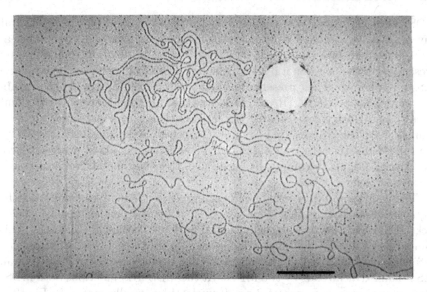

**Fig. 2.5.** Electron micrograph of a segment of purified micronuclear DNA. The circular structure is a defect in the invisible membrane on which the DNA rests. Bar=1 $\mu$m. Courtesy of Gopal Murti

The genes in the micronuclear DNA of stichotrichs have extraordinary structural features not observed in the genes of any other kind of organism. The genes are interrupted by short noncoding segments of DNA called internal eliminated segments, or IESs. A few genes are interrupted by a single IES; most genes contain many IESs. Figure 2.6 contains a diagram of a micronuclear gene that encodes a protein known as $\beta$TP. It has six IESs, which divide the gene into seven segments, known as macronuclear destined segments, or

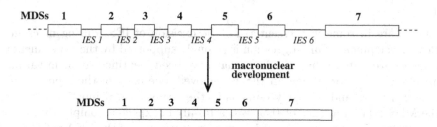

**Fig. 2.6.** A diagram of the arrangement of the six IESs and seven MDSs in the micronuclear gene encoding $\beta$TP protein. During macronuclear development the IESs are excised and the MDSs are ligated (by overlapping of the ends) to yield a macronuclear gene. MDSs are *rectangles*. IESs are *line segments* between *rectangles*.

MDSs. IESs are usually less than 100 nucleotides long, and they may occur at any position in a gene. Extrapolation from the structure of 12 micronuclear genes known in one stichotrich indicates that more than 100,000 IESs are present in all the genes in a haploid genome. The significance of these bizarre IESs is not known. Their presence disables a gene so that it would produce abnormal mRNA products. (Recall, however, that micronuclear genes are silent in any case.) During development of a micronucleus into a macronucleus more than 100,000 IESs are excised from all the genes. Simultaneously, the MDSs are spliced together (by overlapping of the ends, as explained below) to provide transcriptionally competent genes that then become active (expressed) in the new macronucleus.

The lengths and positions of IESs are precisely defined by comparing the sequence of nucleotides in the macronuclear copy of a particular gene (which lacks IESs) with its micronuclear counterpart, or precursor (Fig. 2.7). This comparison not only identifies an IES, but also provides a clue about how an IES is precisely excised during macronuclear development. By definition an IES is flanked at its ends by two MDSs. At the MDS–IES junctions both MDSs contain the same sequence of 3 to 20 nucleotides, as illustrated in Fig. 2.7. This pair of repeats can be any sequence using A, T, G, and C; note that the repeated sequence in Fig. 2.7 is $\frac{\text{ACATTTC}}{\text{TGTAAAG}}$. This suggests that pairs of

**Fig. 2.7.** A section of the micronuclear gene encoding βTP in *Sterkiella histriomuscorum* containing the end portion of MDS 2, IES 2, and the beginning portion of MDS 3. The two repeat sequences (pointers) are *underlined*; the IES is in *lower case* letters.

repeats are in some way essential for the excision of IESs during macronuclear development. This suggestion is strongly supported by the experimental evidence from more than 200 IESs that have been identified so far in various micronuclear genes. When an IES is removed, one copy of the repeat is removed with it, and one copy remains at the newly formed junction between the MDSs. The presence of IESs was a totally unexpected complexity in micronuclear gene structure, requiring massive processing of the DNA to produce macronuclear genes. However, the complexity of micronuclear gene structure and processing extends even beyond the arrangement shown in Fig. 2.6. In the majority of genes the MDSs are in the proper (orthodox) order as shown in Fig. 2.6, and simple excision of the IESs and splicing of the MDSs produces a functional macronuclear gene. Note that Fig. 2.6 gives a somewhat simplified presentation of the MDS splicing as this will become clear later in this section.

However, in some genes the MDSs are in the wrong order, i.e., they are scrambled. Figure 2.8 contains a diagram of the micronuclear gene that encodes a protein called actin in the stichotrich *Sterkiella nova*. The gene is interrupted by eight IESs, creating nine MDSs. The orthodox order of these nine MDSs would be 1-2-3-4-5-6-7-8-9, as defined by the order of MDSs in the macronu-

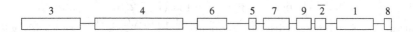

**Fig. 2.8.** Structure of the micronuclear gene-encoding actin protein in the stichotrich *Sterkiella nova*. The nine MDSs are in a scrambled disorder.

clear copy of the gene. However, in the micronuclear version of the actin gene the MDSs are in the scrambled order: 3-4-6-5-7-9-2-1-8. Moreover, MDS 2 is inverted, meaning that the coding sequences of nucleotides in all of the MDSs except for MDS 2 are read from left to right, thereby prescribing the correct order of amino acids in the actin protein. MDS 2 has been inverted so that its coding sequence reads from right to left, and the sense strand of MDS 2 is in the opposite strand than in the other eight MDSs (Fig. 2.8). The processing of this gene during macronuclear development not only requires excision of IESs but also the MDSs must be unscrambled and spliced in the orthodox order. This includes the inverted MDS 2 so that it reads from left to right, i.e., in the same direction as the other eight MDSs. Again, pairs of repeat sequences are present at the ends of MDSs that indicate how MDSs must be joined to create the orthodox order. For example, Fig. 2.9 shows the scrambled MDSs 4-6-5-7

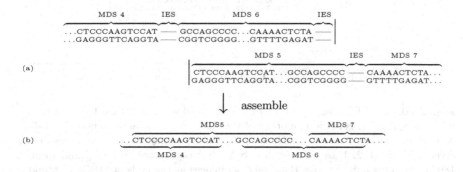

**Fig. 2.9. a** Repeat sequences in MDSs 4-6-5-7 in the micronuclear actin gene. Note the matching of the repeats between the *right end* of MDS 4 and the *left end* of MDS 5, between the *right end* of MDS 5 and the *left end* of MDS 6, and between the *right end* of MDS 6 and the *left end* of MDS 7. **b** The matchings facilitate joining of these four MDSs in the orthodox order, 4-5-6-7. One copy of each repeat is excised with the IESs.

with the sequences at their ends. MDS 4 ends in $\frac{\text{CTCCCAAGTCCAT}}{\text{GAGGGTTCAGGTA}}$, which is

repeated at the beginning of MDS 5. MDS 5 has the sequence $\frac{\text{GCCAGCCCC}}{\text{CGGTCGGGG}}$

at its end, which matches the sequence at the beginning of MDS 6, and finally

MDS 6 ends in $\frac{\text{CAAAACTCTA}}{\text{GTTTTGAGAT}}$, which is present in the beginning of MDS 7.

MDS 2 is a special case because it is inverted (Fig. 2.10). In its micronuclear

position MDS 2 begins with the sequence $\frac{\text{CTTGACGACTCC}}{\text{GAACTGCTGAGG}}$ and ends with

$\frac{\text{ATGTGTAGTAAG}}{\text{TACACATCATTC}}$. Inversion puts the end sequence at the beginning and vice

versa. However, because the DNA strands have polarity, MDS 2 must be rotated 180° on its long axis, which means that the sequence at the beginning

of MDS 2 becomes $\frac{\text{CTTACTACACAT}}{\text{GAATGATGTGTA}}$. Similarly, the sequence at the end of

MDS 2 becomes $\frac{\text{GGAGTCGTCAAG}}{\text{CCTCAGCAGTTC}}$. These two sequences and the beginning

and end of the MDS 2 (after it has been reinverted) match sequences at the
end of MDS 1 and the beginning of MDS 3, respectively. Thus, the rule that
MDS splicing in the orthodox order is guided by repeat pairs is obeyed.

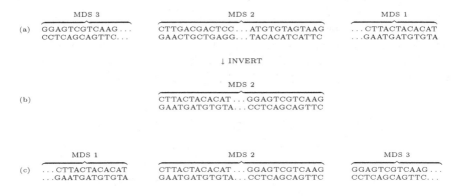

**Fig. 2.10. a–c.** Reinversion of MDS 2 and reordering of MDSs 1 and 3 in the
micronuclear actin gene: **a** The order of the 9 MDSs in the actin gene is 3-4-6-5-7-
9-2-1-8, i.e., MDSs are in the order 3, 2, 1 within the gene. Repeat sequences at the
ends of MDSs 3, 2, 1 are written out. **b** MDS 2 is inverted (reversed and rotated
180° on its long axis). **c** Now the repeat sequences at the ends of MDSs 1, 2, and 3
match.

In another micronuclear gene, encoding the protein αTP, the 14 MDSs are
in the scrambled disorder 1-3-5-7-9-11-2-4-6-8-10-12-13-14 (Fig. 2.11). Note
that here the MDSs fall into a regular pattern of odd-numbered and even-
numbered series, with three nonscrambled MDSs (12-13-14) at the end. In

addition, none of the MDSs is inverted, as is the case of MDS 2 in the actin gene.

**Fig. 2.11.** Structure of the micronuclear gene encoding αTP protein in the stichotrich *Sterkiella nova.*

The most complex pattern of MDS scrambling occurs in the gene that encodes the protein DNA polymerase (the enzyme that synthesizes DNA). In the stichotrich *Sterkiella nova*, there are 45 MDSs (Fig. 2.12), occurring in two unconnected groups in the DNA polymerase gene in the following order: MDS group 1 is 27-26-24-22-20-18-16-14-12-10-8-6-4-1-2-3-5-7-9-11-13-15-17-19-21-23-25-28-30-32-34-36-38-40-42-44-45, and MDS group 2 is 29-31-33-35-37-39-41-43. Moreover, MDSs 27 through 4 in Fig. 2.12 are inverted relative to MDSs 1 through 45. Each time a stichotrich mates it must accomplish the momentous task, and without a single error, of excision of 43 IESs, inverting and reordering MDSs, and splicing the MDSs in the two groups into a single group with the orthodox order 1 through 45.

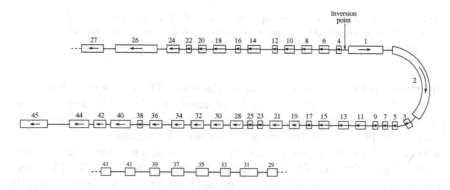

**Fig. 2.12.** Structure of the micronuclear gene encoding the DNA polymerase protein in the stichotrich *Sterkiella nova*. The gene exists in a large segment and in a small segment. Nearly half of the large segment is inverted.

One final, dramatic step in DNA processing occurs after IESs have been removed and MDSs ligated in the orthodox order. About 25,000 genes in the chromosomes in the developing macronucleus are excised from their chromosomal locations, and all the spacer DNA is destroyed. This leaves only very short DNA molecules, each molecule containing a single gene (an occasional molecule may contain two genes). A few nucleotides of a specific sequence are then added to the ends of the DNA strands of these short molecules,

**Fig. 2.13.** An electron micrograph of some DNA molecules isolated from the macronucleus of a stichotrich. Each molecule encodes a single gene. Bar=1$\mu$m. Courtesy of Gopal Murti

and these nucleotides serve to protect the ends. The creation of gene-sized DNA molecules is an extraordinary event, known to occur only in stichotrichs and a few other ciliate groups. Each of the about 25,000 different gene-sized DNA molecules is replicated many times until it is present in hundreds to thousands of identical copies, depending on the species. Thus, the mature macronucleus contains millions of short DNA molecules (Fig. 2.13). The presence of many copies of every gene in the macronucleus allows the organism to produce mRNA molecules, and, in turn, protein molecules, at extraordinary rates. This greatly increases the potential for rapid growth and cell reproduction — two important activities in competition for survival of a species.

Although IESs have been detected in micronuclear genes in two distantly related ciliate species, *Tetrahymena* and *Paramecium*, IESs are far more numerous in stichotrichs. In addition, the gene scrambling phenomenon has not been found in any other ciliate group. Stichotrichs are among the ciliates that have evolved most recently. Thus, micronuclear gene scrambling and a high abundance of IESs must be recent evolutionary innovations. Indeed, a study of the micronuclear actin gene in a series of stichotrichs has shown that IESs are progressively added to the actin gene and the gene becomes scrambled in an increasingly complex way as evolution proceeds.

It may immediately be asked: What are the role(s) of IESs and of MDS scrambling in the life of a stichotrich? It is postulated that both phenomena in some way facilitate the evolution of stichotrichs. Evolution of all organisms, bacteria and eukaryotes alike, proceeds by a change in DNA, i.e., by mutation. Mutations are caused by mutagenic chemicals and radiation, and by mistakes cells make in replicating DNA. A vast majority of mutations are deleterious to the organism, decreasing the organism's ability to survive. On rare occasions a mutation in DNA may improve the function of the product (a protein) encoded by a gene, or even lead to creation of a new, useful gene. This concept is elegantly summarized in Lewis Thomas's statement, "The capacity to blunder slightly is the real marvel of DNA. Without this special attribute, we would still be anaerobic bacteria and there would be no music." Thus, evolution depends on DNA changes, but changes need to be limited because most of them are detrimental. Stated another way, the ratio between unfavorable and favorable mutations severely restricts the evolution of a species. The phenomena of IESs and gene scrambling may alter this ratio in a way that is not yet clear, allowing additional favorable changes in DNA without increasing deleterious changes. This would facilitate evolution of new genes that could improve the function and survivability of a species.

## Notes on References

The study of DNA in stichotrichs began in 1971 with the chance discovery that DNA in the macronucleus is in very short molecules — the shortest DNA molecules occurring in nature. The discovery is described in Prescott et al. [49]. Subsequent discoveries on stichotrich macronuclear DNA are reviewed in Prescott and Rozenberg [54]. A detailed review of ciliate DNA can be found in Prescott [47]. A succinct recent review of DNA processing in stichotrichs can be found in Prescott [48].

# 3

# Molecular Operations for Gene Assembly

We can only speculate about the significance of the extraordinary modifications in the structure of micronuclear genes in stichotrichous ciliates that have occurred during evolution. However, it is clear the presence of IESs and MDSs, and particularly the scrambling of MDSs in micronuclear genes, requires spectacular manipulations of DNA sequences whenever cells mate and then develop new macronuclei. These DNA manipulations are massive, consisting of the excision of more than 100,000 IESs and reordering of many thousands of MDSs. Major progress in understanding this intricate DNA processing has been made by studying the nucleotide sequences in micronuclear genes. The excision of IESs and the unscrambling and joining of MDSs are referred to as gene assembly. In this chapter we postulate three molecular operations based on the pairs of repeats in the ends of MDSs to accomplish the assembling of genes after cell mating. In the description of these three operations the pairs of repeats are called pointers because they point the way for MDS joining.

## 3.1 Homologous Recombination

The assembly of genes during macronuclear formation in stichotrichs takes place by homologous recombination between pointers. Homologous recombination occurs between two DNA molecules (designated 1 and 2 in Fig. 3.1) that possess a segment containing an identical sequence of base pairs. These identical sequences are aligned side-by-side as shown in Fig. 3.1a. Exactly how the two identical segments recognize each other and then align is not known, but it occurs commonly in cells. Once the identical segments are aligned, an enzyme introduces staggered cuts (at the four arrows in Fig. 3.1b) in the same position in each DNA molecule. The staggered cuts create identical single-stranded overhangs (in this case, six nucleotides long, although they can be longer or shorter in other cases). Molecules 1 and 2 now switch parts with each other, and an enzyme called ligase repairs the cuts as shown in Fig. 3.1c.

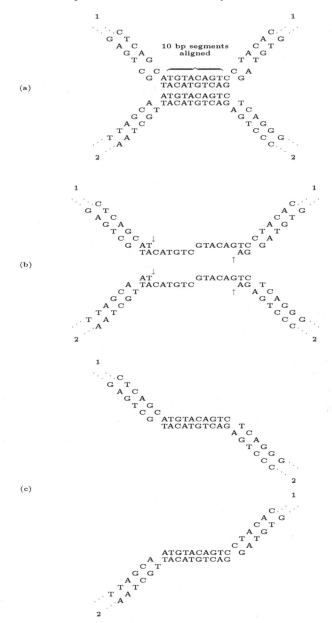

**Fig. 3.1. a–c.** Homologous recombination: **a** Two DNA molecules (1 and 2) align along a 10-bp segment of identical sequence. Other than the aligned segment the two molecules are composed of different sequences. **b** Staggered cuts are made in the two molecules (at the *four arrows*), creating 6-bp, single-stranded overhangs in both molecules. **c** The two molecules recombine in a switch configuration so that the *left part* of molecule 1 is joined to the *right part* of molecule 2, and vice versa.

The result is that the two DNA molecules have switched parts, i.e., they have recombined.

## 3.2 Three Molecular Operations

The processing of DNA requires molecular complexes that perform molecular operations. Molecular operations that carry out gene assembly depend on the MDS–IES structure of micronuclear and intermediate forms of genes. We consider three molecular operations that account for the process of gene assembly.

### Loop Recombination

Loop recombination is a simple operation in which the pair of repeat sequences (which we will call pointers) flanking an IES between two MDSs guides the excision of the IES, with the concomitant joining of the two MDSs. This is illustrated in Fig. 3.2 for the IES between MDS 2 and MDS 3 in the $\beta$TP gene in the ciliate *Sterkiella histriomuscorum* (see Fig. 2.7). In this pair of pointers, P3, the copy of P3 at the end of MDS 2 is the outgoing pointer (of MDS 2), and the copy of P3 at the beginning of MDS 3 is the incoming pointer (of MDS 3). The two copies are aligned in parallel by forming a loop in the DNA molecule. The two copies of P3 are recombined by homologous recombination, as shown in Fig. 3.2. This recombination in the $\beta$TP gene joins MDS 2 to MDS 3 through one recombined copy of P3 and excises the IES as a circular molecule containing the other recombined copy of P3. MDS 2 and MDS 3 have now formed a composite MDS, and the excised IES is destroyed by the cell. Excision of DNA segments as circular molecules is well known in other situations of cell behavior.

The technical name for this operation (taking into account the kind of fold and the kind of repeat) is *(loop, direct repeat) excision*, or *ld* for short.

### Hairpin Recombination

Hairpin recombination is applicable to a portion of a molecule containing a pair of pointers in which one pointer is an inversion of the other. For example, if one copy in a pair of pointers were $\begin{smallmatrix} 5' \text{ CCTGA } 3' \\ 3' \text{ GGACT } 5' \end{smallmatrix}$ , then the other copy would be an inversion, i.e., $\begin{smallmatrix} 5' \text{ TCAGG } 3' \\ 3' \text{ AGTCC } 5' \end{smallmatrix}$ . This is formally illustrated for the segment of the actin gene containing $\overleftarrow{\text{MDS 2}}$ – IES 7 – $\overrightarrow{\text{MDS 1}}$ – IES 8, in which MDS 2 is inverted relative to MDS 1, as indicated by a reverse arrow over MDS 2 pointing in opposite directions (Fig. 3.3). In the unscrambling of the actin gene (Fig. 2.8), MDS 1 will join with MDS 2 through one copy of the pointer P2. This is accomplished by aligning the outgoing copy of pointer P2 of MDS 1

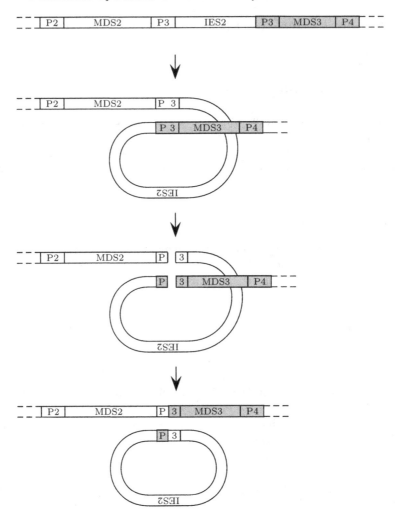

**Fig. 3.2.** Loop recombination. The DNA molecule folds into a simple loop that brings the pair of pointers (*P3*) into side-by-side alignment. Homologous recombination between the two pointers joins MDS 2 with MDS 3 and removes the IES as a closed circle.

with the incoming copy of pointer P2 of MDS 2, followed by recombinational switching between the two copies of the pointer. However, MDS 2 and its incoming pointer P2 have the inverted polarity. By forming a hairpin in the DNA, the two copies of P2 are put into the same polarity. This allows them to undergo homologous recombination, forming the composite MDS from MDS 1 and MDS 2, with one copy of P2 between them. The other recombined copy of P2 becomes part of the excised IES 7, which has become joined to IES 8.

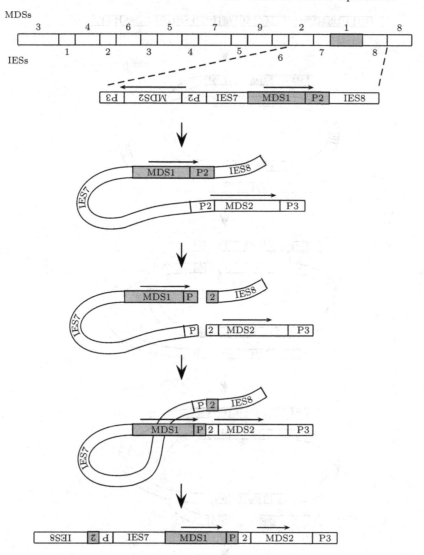

**Fig. 3.3.** Hairpin recombination. The pair of pointers P2 are in opposite orientations or polarities. The molecule folds so as to bring the two pointers into the same polarity. Homologous recombination between pointers P2 now joins MDS 1 with MDS 2 and IES 6 with IES 7.

This composite IES has been shifted out of the actin gene, and it now flanks MDS 1. Subsequently, the actin gene is excised from its chromosomal location by cutting the DNA between MDS 1 and the composite IES 7 – IES 8 and between MDS 9 and its flanking DNA sequence.

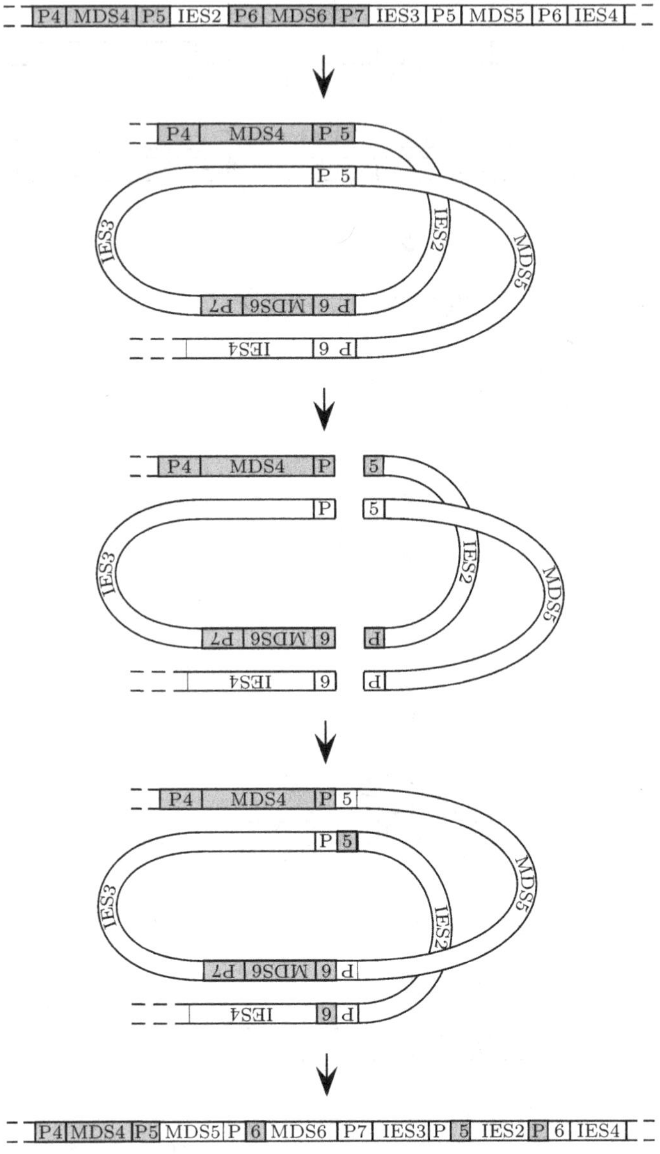

**Fig. 3.4.** Double-loop recombination. The operation of double-loop applies when one pair of pointers (in this case P5) encompasses a segment of a gene that overlaps with a segment encompassed by another pair of pointers (in this case P6). Folding into a double-loop, followed by recombination of P5 with P5 and P6 with P6 rearranges MDSs 4, 5, and 6 in the proper numerical order.

The technical name for this operation is *(hairpin, inverted repeat) excision/reinsertion,* or *hi* for short. Recombination through hairpin formation in double helices of DNA is not well documented experimentally, but it is conceptually quite reasonable as a particular form of homologous recombination.

## Double-Loop Recombination

Double-loop recombination applies to a region of a DNA molecule that contains two pairs of pointers in which the segment of the molecule encompassed by one pair of pointers overlaps with the segment encompassed by the other pair of pointers. This is shown in Fig. 3.4 for a segment of the micronuclear actin gene in *Sterkiella nova,* see Fig. 2.8. The double-folded molecule in Fig. 3.4 satisfies the key requirement of the definition of the double-loop recombination because two pairs of pointers, P5 and P6, are interdigitated, i.e., in the order P5, P6, P7, P5, P6. The molecule folds into two loops so that the two copies of P5 align with each other in one loop, and the two copies of P6 align with each other in the other loop. Thus, the molecule is in position for two recombinations. Recombination between the two copies of P5 joins MDS 4 to MDS 5 through a copy of P5, making the composite MDS 4 – MDS 5. Recombination between the two copies of P6 joins MDS 5 into a composite MDS with MDS 6 through one copy of P6. As a result of these two recombinations IESs 3, 2, and 4 are joined by the other copy of P5 and the other copy of P6. Thus, the three IESs form a composite segment that now resides after MDS 6.

The technical name for this operation is *(double-loop, alternating direct repeat) excision/reinsertion,* or *dlad* for short.

The three operations, loop recombination, hairpin recombination, and double-loop recombination, may be applied in various combinations when necessary to assemble a gene. Moreover, multiple applications of the same or different operations may take place simultaneously or in succession. The process of gene assembly by these operations is a process of pointer removal. When pairs of pointers have been disrupted, with one copy of each removed, a macronuclear version of a gene is formed, free of IESs and pairs of pointers. Finally, the three operations account for all MDS/IES patterns observed thus far in scrambled and nonscrambled genes. The three operations, working singly or in various combinations, account for the excision of IESs and the joining of MDSs in the orthodox order in all nonscrambled and scrambled gene patterns observed experimentally thus far.

## Notes on References

The *ld, hi,* and *dlad* operations are defined in detail in Prescott et al. [51].

# Part II

Formal Modelling of Gene Assembly

# 4

# Model Forming

In the second part of the book we shall formalize the process of gene assembly in ciliates. Such a task requires two steps: First, we need to formalize the objects that are processed, i.e., DNA molecules that constitute micronuclear and intermediate genes. (An *intermediate gene* is obtained during the process of assembling the macronuclear gene from its micronuclear form.) Second, we need to formalize the way that these objects are processed, i.e., we need to formalize the molecular operations used in the gene assembly as well as the effect of these operations on genes.

The formal definitions of the notions that are summarized in this chapter are presented in the following chapters of Part II.

## 4.1 Formalizing Genes

Let us first consider possible formalizations of DNA molecules. A DNA molecule (see Chap. 1) may be seen as a sequence of bases, with four types of bases available: A, C, G, and T. Therefore, one can represent a single-stranded molecule as a string over the alphabet $\mathcal{N} = \{A, C, G, T\}$ of bases, and then a double-stranded molecule as a "double string" over $\mathcal{N}$ (see, e.g., Fig. 1.5). However, this representation is much too detailed for our considerations.

Therefore, rather than considering DNA molecules as double strings over $\mathcal{N}$, in investigation of gene assembly, where the MDS/IES structure of micronuclear and intermediate genes is so crucial, we can use the pictorial *block representation* of genes, as done, e.g., in Figs. 2.6, 2.8, and 2.11.

In the formal context, it is more convenient to denote the $i$th MDS by $M_i$, and its inversion by $\overline{M}_i$. Also, $M_{i,j}$ will denote the composite MDS made up from the MDSs $i$ through $j$. Thus, for instance, Fig. 4.1a represents the MDS/IES structure of a micronuclear gene, while Fig. 4.1b represents the MDS/IES structure of an intermediate gene produced during gene assembly from the gene represented in Fig. 4.1a.

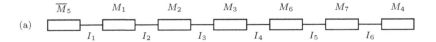

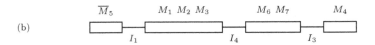

**Fig. 4.1. a** A block representation of an MDS/IES structure consisting of the MDSs $M_1, \ldots, M_7$ and the IESs $I_1, \ldots, I_6$. MDS $M_5$ is inverted. **b** A block representation of an intermediate MDS/IES structure that is obtained from the MDS/IES structure in (a) in the process of gene assembly by splicing MDSs $M_1, M_2,$ and $M_3$ as well as $M_6$ and $M_7$.

Although the block representation is a good visualization of the structures we consider, it is rather cumbersome to be used as a formal tool for reasoning about gene assembly. For this reason, the next abstraction step in our formalization process is to express the block representation as strings over a suitable alphabet. We shall use a symbolic alphabet to denote MDSs and composite MDSs, and consequently genes will be represented as strings over this alphabet. Moreover, in this translation step the identity of IESs will be "forgotten" – we shall have no symbols to denote different IESs. Thus, e.g., the micronuclear MDS/IES structure from Fig. 4.1a will be represented by the string $\overline{M}_5 M_1 M_2 M_3 M_6 M_7 M_4$, while the intermediate MDS/IES structure from Fig. 4.1b will be represented by the string $\overline{M}_5 M_{1,3} M_{6,7} M_4$. These string representations are called **MDS arrangements**.

As explained in Chap. 3, it is the pointer structure of MDSs that is crucial for gene assembly. More specifically, the process of gene assembly is guided by the incoming/outgoing pointers of MDSs. The crucial "guiding feature" is the fact that the outgoing pointer of the MDS $M_n$ is equal to the incoming pointer of the MDS $M_{n+1}$.

Thus, in the investigation of gene assembly one may abstract the structure of each MDS $M_i$ into just an ordered pair $(i, i+1)$ indicating its incoming and outgoing pointers. Following this abstraction step, we move from MDS arrangements to **MDS descriptors**, which are strings of pairs of symbols representing pointers – although certain pairs involve the **beginning marker** $b$ and the **end marker** $e$ that are set to mark the beginning and the ending of the micronuclear gene. For example, the micronuclear MDS arrangement $\overline{M}_5 M_1 M_2 M_3 M_6 M_7 M_4$ becomes the micronuclear MDS descriptor $(\overline{6}, \overline{5})(b, 2)(2, 3)(3, 4)(6, 7)(7, e)(4, 5)$, and the intermediate MDS descriptor $\overline{M}_5 M_{1,3} M_{6,7} M_4$ becomes the intermediate MDS descriptor $(\overline{6}, \overline{5})(b, 4)(6, e)(4, 5)$.

It is often more convenient to deal with strings of pointers rather than with strings made up of ordered pairs of pointers (or markers). Therefore, the next natural simplification of our representation of genes is to use strings consisting of pointers only. Each such string is a "pointer snapshot" of a gene at a given stage of assembly (a snapshot of a gene with only pointers visible). In this way we obtain the framework of **legal strings**. For instance, the micronuclear MDS descriptor $(\overline{6}, \overline{5})(b, 2)(2, 3)(3, 4)(6, 7)(7, e)(4, 5)$ is translated in this way into the legal string $\overline{6}\,\overline{5}\,2\,2\,3\,3\,4\,6\,7\,7\,4\,5$, and the intermediate MDS descriptor $(\overline{6}, \overline{5})(b, 4)(6, e)(4, 5)$ is translated into the legal string $\overline{6}\,\overline{5}\,4\,6\,4\,5$.

By inspecting the MDS/IES structure of micronuclear or intermediate genes one may notice various differences in their complexity. Thus, e.g., the structure of the micronuclear gene encoding the $\beta$TP protein in Fig. 2.6 is very simple. This simplicity is well reflected in the "orthodox" structure of the corresponding legal string

$$2\,2\,3\,3\,4\,4\,5\,5\,6\,6\,7\,7, \tag{4.1}$$

where there are no overlaps between any two different (pairs of) pointers.

The situation is different if one considers the partial MDS/IES structure in Fig. 4.2. Applying the *hi*-rule (see Chap. 3) to pointer P4 yields, as illustrated in Fig. 4.3a, the MDS/IES structure represented in Fig. 4.3b, while applying the *hi*-rule to pointer P5 yields the MDS/IES structure given in Fig. 4.4b as illustrated in Fig. 4.4a. These two application of the *hi*-rule interfere with each other. That is, they cannot be applied independently: after the *hi*-rule is applied to pointer P4, it can no longer be applied to pointer P5, and, the other way round, after the *hi*-rule is applied to pointer P5, it can no longer be applied to P4. This interdependence is well seen in the structure of the legal string corresponding to the MDS/IES structure from Fig. 4.2:

$$\ldots 7\,8\,3\,4\,6\,7\,\overline{5}\,\overline{4}\,5\,6 \ldots$$

where the inverted repeat of pointer P5 overlaps with the inverted repeat of pointer P4, as illustrated below:

$$\ldots 7\,8\,3\,4\,6\,7\,\overline{5}\,\overline{4}\,5\,6 \ldots$$

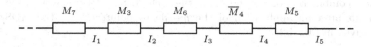

**Fig. 4.2.** The block representation of a partial MDS/IES structure consisting of the MDSs $M_3, M_4, M_5, M_6$, and $M_7$ and the IESs $I_1, \ldots, I_5$. The MDS $M_4$ is inverted (indicated in the figure by a *bar* above $M_4$).

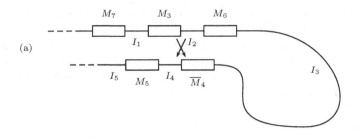

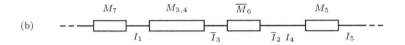

**Fig. 4.3. a** An illustration for the application of the *hi*-rule to pointer P4, which guides the recombination of $M_3$ and $M_4$ in the MDS/IES structure from Fig. 4.2. This operation splices together MDSs $M_3$ and $M_4$ by folding the molecule in a hairpin formation. **b** The result after an application of the *hi*-rule to pointer P4 in the MDS/IES structure given in Fig. 4.2.

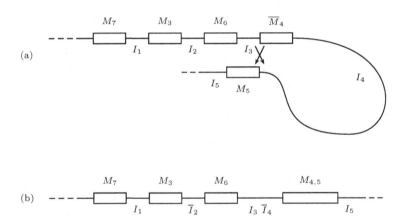

**Fig. 4.4. a** An illustration for the application of the *hi*-rule to pointer P5, which guides the recombination of $M_4$ and $M_5$ in the MDS/IES structure given in Fig. 4.2. **b** The result after an application of the *hi*-rule to pointer P5 in the MDS/IES structure given in Fig. 4.2.

These considerations lead one to the observation that perhaps essential information needed for performing gene assembly is included in the overlap structure of pointers. This overlap structure is represented by **overlap graphs**. For instance, the overlap graph for the MDS/IES structure from

Fig. 2.6 is given in Fig. 4.5, while (part of) the overlap graph for the MDS/IES structure from Fig. 4.2 is given in Fig. 4.6. In these figures each vertex (drawn as a circle) represents (both occurrences of) a pointer, and the presence of an edge between two pointers indicates that the two pointers overlap. Moreover, a minus sign of a vertex indicates that the vertex has a direct repeat, while a plus sign indicates inverted repeat. Following this line of reasoning, we formulate the MDS/IES structure through its overlap graph.

**Fig. 4.5.** The overlap graph for the MDS/IES structure from Fig. 2.6. Each vertex (drawn as a *circle*) represents a pointer together with its qualification (positive/negative). In this example all vertices have a minus sign because all pointers are negative. Moreover, there are no edges in the graph because the pointers in the legal string (4.1) do not overlap.

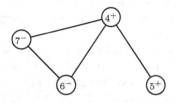

**Fig. 4.6.** The overlap graph for the partial MDS/IES structure from Fig. 4.2. The legal string corresponding to this part is $7\,8\,3\,4\,6\,7\,\overline{5}\,\overline{4}\,5\,6$, where pointers P4 and P5 are positive, and all other pointers are negative (this is reflected by the signs of the vertices). Also, e.g., there is an edge between vertices 6 and 7 in the overlap graph, because pointers P7 and P6 overlap.

## 4.2 Levels of Abstraction

Since we know the base sequence for each MDS and IES, translating from the base sequence representation of genes into their block representation is one-to-one. In other words, these representations are mutually translatable, i.e., given one representation the other one can be unambiguously recovered. Thus, in this translation no information is lost, and therefore we say that both representations are on the same abstraction level.

On the other hand, in translating from block representations to MDS arrangements we lose the information about the identity of IESs. For instance, for the MDS arrangement $\overline{M}_5 M_{1,3} M_{6,7} M_4$, which was obtained from the MDS/IES structure from Fig. 4.1b, we know that there is an IES between $M_{1,3}$ and $M_{6,7}$, and that there is an IES between $M_{6,7}$ and $M_4$, but we have no explicit information available that can tell us the difference in identity of these IESs. Such explicit information is available in the block representation. Therefore, in going from block representations to MDS arrangements we proceed to a model on a higher level of abstraction.

Translating from MDS arrangements to MDS descriptors preserves the abstraction level because given one of the two representations of a gene the other one can be unambiguously recovered. For example, the MDS arrangement $\overline{M}_5 M_{1,3} M_{6,7} M_4$ is translated letter-by-letter into the corresponding MDS descriptor $(\overline{6},\overline{5})(b,4)(6,e)(4,5)$, where $M_5$ corresponds to $(\overline{6},\overline{5})$, $M_{1,3}$ corresponds to $(b,4)$, etc. The same holds in the other direction, because we know that $(\overline{6},\overline{5})$ corresponds to $\overline{M}_5$, $(b,4)$ corresponds to $M_{1,3}$, etc.

When moving from MDS descriptors to legal strings, we again lose information: translation from an MDS descriptor into the corresponding legal string is unique, but the same legal string may be obtained from different MDS descriptors. Hence, e.g., the MDS descriptor $(b,2)(3,4)(5,e)(\overline{5},\overline{4})(\overline{3},\overline{2})$ is translated into the legal string $2\,3\,4\,5\,\overline{5}\,\overline{4}\,\overline{3}\,\overline{2}$, but also the MDS descriptor $(2,3)(4,5)(\overline{e},\overline{5})(\overline{4},\overline{3})(\overline{2},\overline{b})$ is translated into the same legal string. Therefore, in translating from MDS descriptors into legal strings we proceed to a model of a higher abstraction level.

Also, the same holds for translating from legal strings to overlap graphs. In this translation the linear structure of a legal string is lost, and so many legal strings yield the same overlap graph. For example, the legal strings $6\,5\,\overline{7}\,4\,4\,6\,\overline{5}\,2\,3\,3\,2\,7$, $7\,5\,6\,\overline{7}\,\overline{5}\,7\,4\,3\,3\,4\,2\,2$, $6\,5\,\overline{7}\,4\,4\,6\,\overline{5}\,2\,3\,3\,2\,7$, and many others yield the overlap graph in Fig. 4.7.

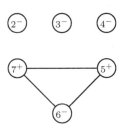

**Fig. 4.7.** The overlap graph that is obtained from many different legal strings, where the pointers P2, P3, P4, and P6 are negative, and pointers P5 and P7 are positive.

In the following chapters we shall study in detail the models of MDS descriptors, legals strings, and overlap graphs. We study them in this order, and hence in the increasing order of abstraction.

## 4.3 Formalizing Molecular Operations

Whichever model we choose for representing gene structures, we also have to represent the molecular operations from Chap. 3 within this model.

Consider, e.g., the gene encoding of the $\beta$TP protein from Fig. 2.6. The corresponding MDS descriptor is $(b, 2)(2, 3)(3, 4)(4, 5)(6, 7)(7, e)$, the corresponding legal string is $2\,2\,3\,3\,4\,4\,5\,5\,6\,6\,7\,7$, and the overlap graph is given in Fig. 4.5.

Applying $ld$-rule to remove IES 2 yields the gene with the block representation given in Fig. 4.8.

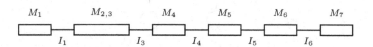

**Fig. 4.8.** The block representation of the gene represented in Fig. 2.6 after the $hi$-rule has been applied to pointer P3 to remove IES 2 and splice $M_2$ and $M_3$.

In the model of MDS descriptors this instance of the $ld$-rule can be formalized as

$$(2, 3)(3, 4) \longrightarrow (2, 4),$$

which reads "replace the substring $(2, 3)(3, 4)$ by the substring $(2, 4)$." This rule instance applied to the MDS descriptor $(b, 2)(2, 3)(3, 4)(4, 5)(6, 7)(7, e)$ gives the MDS descriptor $(b, 2)(2, 4)(4, 5)(6, 7)(7, e)$, which, indeed, corresponds to the block representation from Fig. 4.8.

In the model of legal strings this instance of the $ld$-rule can be formalized as

$$3\,3 \longrightarrow \Lambda,$$

which reads "remove the substring 33." Applying this rule to the legal string $2\,2\,3\,3\,4\,4\,5\,5\,6\,6\,7\,7$ gives the legal string $2\,2\,4\,4\,5\,5\,6\,6\,7\,7$, which also corresponds to the block representation from Fig. 4.8.

Finally, on the level of overlap graphs this instance of the $ld$-rule can be formalized as

$$3^- \longrightarrow \emptyset,$$

which reads "remove the negative isolated vertex 3." (A vertex is isolated if it is not connected by an edge to any other vertex.) When this rule is applied to the overlap graph from Fig. 4.5, we obtain the overlap graph of the legal string $2\,2\,4\,4\,5\,5\,6\,6\,7\,7$ acquired from the graph in Fig. 4.5 by deleting the vertex corresponding to pointer 3.

Consider now a gene for which the block representation is given in Fig. 4.9. The corresponding MDS descriptor is $(5, e)(b, 2)(4, 5)(\overline{3}, \overline{2})(3, 4)$, the corresponding legal string is $5\,2\,4\,5\,\overline{3}\,\overline{2}\,3\,4$, and the overlap graph is given in Fig. 4.10.

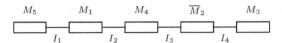

**Fig. 4.9.** The block representation of an MDS/IES structure consisting of the MDSs $M_1, \ldots, M_5$ and the IESs $I_1, I_2, I_3$, and $I_4$. MDS $M_2$ is inverted.

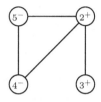

**Fig. 4.10.** The overlap graph of the MDS/IES structure from Fig. 4.9

Applying the *hi*-rule to pointer P2 yields the gene with the block representation given in Fig. 4.11.

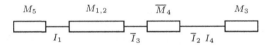

**Fig. 4.11.** The block representation of an MDS/IES structure from Fig. 4.9 after an application of the *hi*-rule to pointer P2

On the level of MDS descriptors this instance of the *hi*-rule can be formalized as the following string replacement rule:

$$(b, 2)(4, 5)(\overline{3}, \overline{2}) \longrightarrow (b, 3)(\overline{5}, \overline{4}),$$

using which one substitutes the substring $(b, 2)(4, 5)(\overline{3}, \overline{2})$ by the substring $(b, 3)(\overline{5}, \overline{4})$.

This rule instance applied to the MDS descriptor $(5, e)(b, 2)(4, 5)(\overline{3}, \overline{2})(3, 4)$ yields the MDS descriptor $(5, e)(b, 3)(\overline{5}, \overline{4})(3, 4)$ which, indeed, corresponds to the block representation from Fig. 4.11.

On the level of legal strings the above instance of the *hi*-rule can be formalized as a string rewriting rule:

$$2\,4\,5\,\overline{3}\,\overline{2} \longrightarrow 3\,\overline{5}\,\overline{4}$$

using which one substitutes the substring $2\,4\,5\,\overline{3}\,\overline{2}$ by the substring $3\,\overline{5}\,\overline{4}$. This rule instance applied to the legal string $5\,2\,4\,5\,\overline{3}\,\overline{2}\,3\,4$ yields the legal string $5\,3\,\overline{5}\,\overline{4}\,3\,4$, which, again, corresponds to the block representation in Fig. 4.11.

Finally, on the level of overlap graphs this instance of the *hi*-rule (applicable to pointer P2) can be formalized as a graph rewriting rule by which

- the sign of each vertex that is connected to 2 is changed,
- for any two different vertices $u$ and $v$ connected to 2, the edge between them is removed if it was there, or the edge between $u$ and $v$ is established if it was not there, and
- the vertex 2 is removed from the graph.

This rule instance applied to the overlap graph from Fig. 4.10 yields the overlap graph from Fig. 4.12, which corresponds to the block representation from Fig. 4.11.

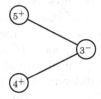

**Fig. 4.12.** The overlap graph obtained from the graph in Fig. 4.10 by using the overlap graph version of the *hi*-rule applied to pointer P2

Also, the application of an instance of a *dlad*-rule can be formalized in each of these three models. However, because it concerns two pairs of pointers, it is somewhat too involved to be discussed in this informal chapter. The reader is referred to Chaps. 7, 9, and 11 for formalizations of the *dlad*-rule in our three models.

## 4.4 Marriage of Models

This book discusses the three basic models of gene assembly: MDS descriptors, legal strings, and overlap graphs, which were informally introduced above. As we have pointed out, proceeding from MDS descriptors to legal strings and then to overlap graphs corresponds to increasing the level of abstraction. Therefore an important question is whether we have not abstracted too much (e.g., when proceeding to the most abstract level of overlap graphs). Since increasing the level of abstraction always means some loss of information, the answer to this question depends on the types of problems that one intends to investigate. A central topic of our research concerns strategies for gene assembly, where strategy is a step-by-step sequence of applications of either a single rule or a group of rules in one step. In particular, we are interested in successful strategies, i.e., those strategies that lead to the assembly of macronuclear genes from their micronuclear forms. It turns out (see Chaps. 9 and 11) that this topic can be equally well investigated in each of our models in the sense that a successful strategy in any of these models can be translated into a successful strategy in any other model. For example, a successful strategy formulated in terms of rewriting rules for MDS descriptors can be directly translated into a successful strategy formulated in terms of rewriting rules for legal strings and the other way around.

Each of these three models has its merits. For instance, working with MDS descriptors is very intuitive, because their form is close to the real MDS/IES structure of genes. However, the *ld-*, *hi-*, and *dlad-*rules for MDS descriptors are quite cumbersome to deal with. These rules for the legals strings turn out to be elegant and also easier to reason about. The form of the overlap graphs no longer resembles MDS/IES structures. However, they are quite convenient, e.g., for formulating possible steps of gene assembly strategies. For example, two vertices from different components of an overlap graph can always be processed in the same step (a component of an overlap graph is a set of vertices that are not connected by an edge to any vertex outside this set).

Also, different models admit different techniques. For instance, in proving properties of gene assembly on the level of legal strings one uses mathematical techniques from combinatorics on words, while in reasoning about graphs one uses techniques of graph theory.

## Notes on References

- This chapter is a preview for the other chapters in Part II, where the three models for gene assembly, representations of micronuclear and intermediate genes, as well as the operations for these representations are defined and investigated in more detail. The main ideas behind this chapter are based on the articles by Prescott et al. [51] and Prescott and Rozenberg [53].

# 5

## Mathematical Preliminaries

In the chapters to follow we shall consider gene assembly from a more formal point of view through operations on strings and graphs. In this chapter we introduce the basic notions and notations concerning strings and graphs needed in those chapters.

### 5.1 Sets and Functions

If $X$ is a finite set, then $\mathrm{card}(X)$ denotes the number of its elements. Let $\mathbb{N}$ denote the set of positive integers, that is, $\mathbb{N} = \{1, 2, \dots\}$. For integers $k, n \in \mathbb{N}$ with $k \leq n$, denote by $[k, n]$ the interval of integers between $k$ and $n$, i.e.,

$$[k, n] = \{k, k+1, \dots, n\}.$$

The empty set is denoted by $\emptyset$.

Let $X$ and $Y$ be two sets. Their **union** and **intersection** are denoted by $X \cup Y$ and $X \cap Y$, respectively. The **difference** of $X$ and $Y$ is the set $X \setminus Y = \{x \mid x \in X \text{ and } x \notin Y\}$. If $Y$ is a subset of $X$, then we write $Y \subseteq X$. The sets $X$ and $Y$ are **disjoint** if $X \cap Y = \emptyset$.

A family $\{X_1, X_2, \dots, X_n\}$ of subsets of a set $X$ is a **partition** of $X$, if $X_i \cap X_j = \emptyset$ for all $i \neq j$ and $\cup_{i=1}^{n} X_i = X$, i.e., if every element of $X$ belongs to a unique set in the family.

A **function** (or a **mapping**) $\varphi \colon X \to Y$ from a set $X$ into a set $Y$ assigns to each element $x \in X$ a unique element $y = \varphi(x)$ of $Y$.

A function $\varphi \colon X \to Y$ is

- **injective** if for all different $x_1, x_2 \in X$, also $\varphi(x_1) \neq \varphi(x_2)$
- **bijection** if $\varphi$ is injective and onto $Y$, i.e., if for each $y \in Y$, there exists a unique $x \in X$ such that $y = \varphi(x)$
- a **permutation** (on $X$) if $\varphi$ is a bijection with $Y = X$

If $\varphi \colon [1, n] \to [1, n]$ is a permutation on the interval $[1, n] = \{1, 2, \dots, n\}$, then we often denote $\varphi$ by $(\varphi(1)\, \varphi(2)\, \dots\, \varphi(n))$.

*Example 5.1.* Let $\varphi\colon [1,5] \to [1,5]$ be a mapping defined by $\varphi(1) = 2$, $\varphi(2) = 1$, $\varphi(3) = 5$, $\varphi(4) = 4$, and $\varphi(5) = 3$. Since $\varphi(i) \neq \varphi(j)$ whenever $i \neq j$, $\varphi$ is a permutation on $[1,5]$. We can then write $\varphi$ in the form $(2\,1\,5\,4\,3)$.    □

If $\varphi\colon X \to Y$ is a bijection, then its **inverse** is the function $\varphi^{-1}\colon Y \to X$ such that $\varphi^{-1}(y) = x$, if $\varphi(x) = y$. It is clear that the inverse function of a bijection is well defined, i.e., it is a function.

An **operation** is a partial mapping $\varphi\colon X \to Y$, where $\varphi$ is defined on a subset of $X$, called the **domain** of $\varphi$ and denoted by $\mathrm{dom}(\varphi)$. Hence $\varphi$ is a function from $\mathrm{dom}(\varphi)$ to $Y$. We also say that an operation $\varphi$ is **applicable** to an element $x \in X$, if $x \in \mathrm{dom}(\varphi)$.

The **composition** of two functions (or operations, resp.) $\varphi_1\colon X \to Y$ and $\varphi_2\colon Y \to Z$ is defined to be $\varphi_2\varphi_1\colon X \to Z$ such that $\varphi_2\varphi_1(x) = \varphi_2(\varphi_1(x))$ for all $x \in X$. Then $\mathrm{dom}(\varphi_2\varphi_1) = \{x \mid x \in \mathrm{dom}(\varphi_1) \text{ and } \varphi_1(x) \in \mathrm{dom}(\varphi_2)\}$.

*Example 5.2.* Let $X = [1,6]$, $Y = [1,5]$, and $Z = [1,4]$. Define the operations $\varphi_1$ with $\mathrm{dom}(\varphi_1) = [2,5]$ and $\varphi_2$ with $\mathrm{dom}(\varphi_2) = [1,4]$ by

$$\varphi_1(2) = 5, \quad \varphi_1(3) = 2, \quad \varphi_1(4) = 3, \quad \varphi_1(5) = 5,$$
$$\varphi_2(1) = 1, \quad \varphi_2(2) = 1, \quad \varphi_2(3) = 4, \quad \varphi_2(4) = 4.$$

Then $\varphi_2\varphi_1(3) = 1$, $\varphi_2\varphi_1(4) = 4$, and $\mathrm{dom}(\varphi_2\varphi_1) = \{3,4\}$.    □

A **rule instance** is an ordered pair $(u,v)$ of objects, usually denoted by $u \to v$ and read "$u$ is replaced by $v$." A **rule** is a set of rule instances.

## 5.2 Strings

Let $\Sigma = \{a_1, a_2, \ldots\}$ be a set of symbols. We call $\Sigma$ an **alphabet**, and its elements are called **letters**. We shall consider both finite and infinite alphabets in this book. A **string over** $\Sigma$ is a finite sequence of letters from $\Sigma$, i.e., each string can be written as $u = a_{i_1} a_{i_2} \ldots a_{i_m}$, where each $a_{i_j}$ is a letter in $\Sigma$. Here $m$ is the **length** of the string $u$. The string of length zero is called the **empty string**, and it is denoted by $\Lambda$. The set of all strings over the alphabet $\Sigma$ is denoted by $\Sigma^*$.

*Example 5.3.* (1)  An alphabet that has two letters is called **binary**. Such an alphabet, e.g., $\Sigma = \{0,1\}$, is often used for basic encoding of structures in computer science. In this alphabet, the strings are $\Lambda, 0, 1, 00, 01, 10, 11, 000, \ldots$, and, for instance, the string $u = 0010011 \in \Sigma^*$ has length seven.

(2)  The alphabet $\{A, T, C, G\}$ is the alphabet of nucleotides. The strings over this alphabet represent single-stranded molecules. To represent double-stranded molecules one can use, for instance, the alphabet consisting of the following 12 letters:

$$\binom{A}{T}, \quad \binom{T}{A}, \quad \binom{C}{G}, \quad \binom{G}{C},$$

$$\binom{A}{\_}, \quad \binom{T}{\_}, \quad \binom{C}{\_}, \quad \binom{G}{\_},$$

$$\binom{\_}{T}, \quad \binom{\_}{A}, \quad \binom{\_}{G}, \quad \binom{\_}{C}.$$

For example, the string $\binom{A}{T}\binom{C}{G}\binom{G}{C}\binom{A}{T}$ represents a DNA molecule, one strand of which is given by ACGA, and the other strand TGCT is complementary to it. On the other hand, the string $\binom{A}{T}\binom{C}{G}\binom{G}{C}\binom{A}{\_}\binom{G}{\_}$ represents a double-stranded molecule, where the upper strand has a suffix AG that has no complementary lower strand. Note, however, that the string $\binom{A}{\_}\binom{\_}{T}$ does not represent a DNA molecule. □

Let $u$ and $v$ be two strings over an alphabet $\Sigma$. We say that $v$ is

- a **substring** of $u$, if $u = w_1 v w_2$ for some strings $w_1, w_2 \in \Sigma^*$. Moreover, a substring $v$ of $u$ is **proper**, if $v$ is nonempty and different from the whole string $u$,
- a **conjugate** of $u$, if $u = w_1 w_2$ and $v = w_2 w_1$ for some strings $w_1$ and $w_2$,
- a **scattered substring** of $u$, if $v$ is obtained from $u$ by deleting some occurrences of letters from $u$, i.e., if $u = u_1 v_1 u_2 v_2 \ldots u_n v_n u_{n+1}$, and $v = v_1 \ldots v_n$ for some strings $v_1, v_2, \ldots, v_n$ and $u_1, u_2, \ldots, u_{n+1}$ with $n \geq 1$. Moreover, $v$ is a **proper** scattered substring of $u$, if $v \neq \Lambda$ and $v \neq u$.

Also, let $u = a_1 a_2 \ldots a_n$, where $a_i \in \Sigma$ for each $i$. Then a string $v$ is called a **permutation** of $u$, if there exists a permutation $(i_1 i_2 \ldots i_n)$ of the index set $[1, n]$ such that $v = a_{i_1} a_{i_2} \ldots a_{i_n}$.

*Example 5.4.* Consider the alphabet $\Sigma = [1, 6]$, and let $u = 242661$ be a string over $\Sigma$ (of length six). Then $v = 426$ is a (proper) substring of $u$, because $u = 2\mathbf{426}61$. Also, $612$ is a substring of the conjugate $661242$ of $u$. Note that $612$ is not a substring of $u$. The string $2266$ is a (proper) scattered substring of $u$, since $u = \mathbf{2}4\mathbf{266}1$. □

## 5.3 Signed Strings

Often in dealing with letters and alphabets it is useful to have two versions for each letter, e.g., $a^{+1}$ and $a^{-1}$ for letter $a$. Usually one version is chosen to be basic, and then the other one is a "signing" of the basic one. Hence one may choose $a^{+1}$ to be the basic version and write it simply as $a$, and then $a^{-1}$ is the signed version of $a$, meaning that the basic version, viz. $a$, becomes decorated (signed) by $-1$. This kind of signing is quite useful when dealing with applications of pointers considered in this book (see Chap. 3). In this context, we shall use the bar above a letter rather than the superscript $-1$ to denote the signed version of a letter; hence we shall write $\bar{a}$ rather than $a^{-1}$.

To be more formal, let $\Sigma$ be an alphabet, and let $\overline{\Sigma} = \{\overline{a} \mid a \in \Sigma\}$ be a signed copy of $\Sigma$ (hence disjoint from $\Sigma$). Then the total alphabet $\Sigma \cup \overline{\Sigma}$ is a **signed alphabet**. The set of all strings over $\Sigma \cup \overline{\Sigma}$ is denoted by

$$\Sigma^{\maltese} = (\Sigma \cup \overline{\Sigma})^*.$$

A string $v \in \Sigma^{\maltese}$ is called a **signed string over** $\Sigma$. We adopt the convention that $\overline{\overline{a}} = a$ for each letter $a \in \Sigma$.

Let $v \in \Sigma^{\maltese}$ be a signed string over $\Sigma$. We say that a letter $a \in \Sigma \cup \overline{\Sigma}$ **occurs** in $v$, if $a$ or $\overline{a}$ is a substring of $v$. Let $\mathrm{dom}(v) \subseteq \Sigma$, called the **domain** of $v$, be the set of the (unsigned) letters that occur in $v$.

*Example 5.5.* Let $\Sigma = \{a, b\}$. Then the signed copy of $\Sigma$ is the alphabet $\overline{\Sigma} = \{\overline{a}, \overline{b}\}$. Now, $\Sigma^*$ is the set of all strings made from $a$ and $b$, and $\Sigma^{\maltese}$ is the set of all strings made from $a$, $b$, $\overline{a}$ and $\overline{b}$. Hence $u = abaa$ is a string over $\Sigma$, while $v = a\overline{b}\overline{a}a$ is a signed string over $\Sigma$. Note that also $u$ is a signed string over $\Sigma$ although it does not use signings.

The letters $a$ and $b$ occur in the signed string $v = a\overline{b}\overline{a}a$, although $b$ is not a substring of $v$ – the signed copy $\overline{b}$ of $b$ is a substring of $v$. The empty string $\Lambda$ belongs to both $\Sigma^*$ and $\Sigma^{\maltese}$.  □

The signing $a \mapsto \overline{a}$ of letters can be extended to longer strings as follows. For a nonempty signed string $u = a_1 a_2 \ldots a_n \in \Sigma^{\maltese}$, where $a_i \in \Sigma \cup \overline{\Sigma}$ for each $i$, let the **inversion** of $u$ be

$$\overline{u} = \overline{a}_n \overline{a}_{n-1} \ldots \overline{a}_1 \in \Sigma^{\maltese}.$$

Moreover, let $u^R = a_n a_{n-1} \ldots a_1$ and $u^C = \overline{a}_1 \overline{a}_2 \ldots \overline{a}_n$ be the **reversal** and the **complement** of $u$, respectively. It is then clear that $\overline{u} = (u^R)^C = (u^C)^R$.

*Example 5.6.* Let $\Sigma = \{a, b\}$ as in Example 5.5. For the string $v = a\overline{b}\overline{a}a$, we have $\overline{v} = \overline{a}ab\overline{a}$ and $(v^C)^R = (\overline{a}ba\overline{a})^R = \overline{a}ab\overline{a} = \overline{v}$.  □

Let $\Sigma$ and $\Gamma$ be two alphabets. A function $\varphi \colon \Sigma^* \to \Gamma^*$ is a **morphism** if $\varphi(uv) = \varphi(u)\varphi(v)$ for all strings $u, v \in \Sigma^*$. A morphism $\varphi \colon \Sigma^{\maltese} \to \Gamma^{\maltese}$ between signed strings is required to satisfy also that $\varphi(\overline{u}) = \overline{\varphi(u)}$ for all $u, v \in \Sigma^{\maltese}$. Note that the images $\varphi(a)$ for the letters $a \in \Sigma$ determine a morphism $\varphi$, that is, if the images of the letters $a \in \Sigma$ are given, then the image of each string is determined.

*Example 5.7.* Let $\Sigma = \{a, b\}$ and $\Gamma = \{0, 1, 2\}$, and let $\varphi \colon \Sigma^{\maltese} \to \Gamma^{\maltese}$ be the morphism defined by $\varphi(a) = 0$ and $\varphi(b) = 2$. Then for $u = a\overline{b}b$, we have $\varphi(u) = 0\overline{2}2$ and $\varphi(\overline{u}) = \varphi(\overline{b}b\overline{a}) = \overline{2}2\overline{0} = \overline{\varphi(u)}$.  □

Two signed strings $u \in \Sigma^{\maltese}$ and $v \in \Gamma^{\maltese}$ are said to be **isomorphic** if there exists an injective morphism $\varphi \colon \Sigma^{\maltese} \to \Gamma^{\maltese}$ with $\varphi(\Sigma) \subseteq \Gamma$ such that $\varphi(u) = v$. Such a morphism $\varphi$ is called an **isomorphism**. Hence two strings are isomorphic if each of the strings can be obtained from the other by renaming letters. Clearly, the "structure" of isomorphic strings is the same.

For each alphabet $\Sigma$, let $\|.\|$ be the morphism from $\Sigma^{\circledast}$ to $\Sigma^*$ that unsigns the letters, i.e., it removes the bars from the letters: for all $a \in \Sigma$,

$$\|a\| = a = \|\bar{a}\|.$$

A signed string $v$ over $\Sigma$ is a **signing** of a string $u \in \Sigma^*$, if $\|v\| = u$. A signing of a permutation of $u$ is said to be a **signed permutation of** $u$.

*Example 5.8.* Let $u = ab\bar{b}ca\bar{c}$ be a signed string over the alphabet $\Sigma = \{a, b, c\}$. Then $u$ is isomorphic to the signed string $v = 2\,3\,\bar{3}\,1\,2\,\bar{1}$ over the alphabet $\Sigma' = [1, 3]$. In this example, the isomorphism $\varphi$ that carries $u$ onto $v$ is defined by the images of the letters of $\Sigma$: $\varphi(a) = 2$, $\varphi(b) = 3$, and $\varphi(c) = 1$. Hence also $\varphi(\bar{a}) = \bar{2}$, $\varphi(\bar{b}) = \bar{3}$, and $\varphi(\bar{c}) = \bar{1}$.

Also, the string $u$ is a signing of the string $w = abbcac$, i.e., $w = \|u\|$. Note that $u$ and $\|u\|$ are *not isomorphic*. □

## 5.4 Circular Strings

Strings can be naturally applied to denote *linear* DNA molecules as indicated in Example 5.3(2). However, in the gene assembly process, one also encounters circular DNA molecules. In this section we introduce circular strings to model such molecules.

We define a relation $\sim$ over $\Sigma^{\circledast}$ in the following way. For two signed strings $u, v$ over $\Sigma$, we say that $u$ is in relation $\sim$ with $v$, denoted by $u \sim v$, if $u$ and $v$ are conjugates, or there is a conjugate $w$ of $v$ such that $u = \bar{w}$.

*Example 5.9.* Let $u_1 = 2\,3\,\bar{4}\,5\,6$, $u_2 = \bar{4}\,5\,6\,2\,3$, $u_3 = \bar{5}\,4\,\bar{3}\,\bar{2}\,\bar{6}$, and $u_4 = 6\,5\,\bar{4}\,3\,2$. Then $u_1$ and $u_2$ are conjugates and so, $u_1 \sim u_2$. Also, since $u_1$ and $6\,2\,3\,\bar{4}\,5$ are conjugates, and $u_3 = \overline{6\,2\,3\,\bar{4}\,5}$, it follows that $u_1 \sim u_3$. However, $u_1$ and $u_4$ are not in the relation $\sim$. Note also that $u_2 \sim u_3$. □

It is straightforward to prove that $\sim$ is an equivalence relation over $\Sigma^{\circledast}$, i.e., for all $u, v, w \in \Sigma^{\circledast}$,

- $u \sim u$ (**reflexivity**)
- if $u \sim v$, then also $v \sim u$ (**symmetry**), and
- if $u \sim v$ and $v \sim w$, then also $u \sim w$ (**transitivity**)

For each $u \in \Sigma^{\circledast}$, the set

$$[u] = \{v \mid v \sim u\}$$

is called a (**signed**) **circular string** over $\Sigma$. Thus, a circular string over $\Sigma$ is an equivalence class of $\sim$. It is clear that the equivalence classes $[u]$, for $u \in \Sigma^{\circledast}$, form a partition of $\Sigma^{\circledast}$. Also, a (noncircular) signed string $v \in \Sigma^{\circledast}$ is sometimes called a **linear string**.

Since $u \sim \bar{u}$, for any string $u$, it follows that $[u] = [\bar{u}]$. Also $[u] = [v]$, for any conjugate $v$ of $u$.

*Example 5.10.* Let $u_1 = 2\,4\,\overline{5}\,3$ and $u_2 = \overline{4}\,\overline{2}\,\overline{3}\,5$. Then $u_1 \sim u_2$ and so, $[u_1] = [u_2]$. Note that $[u_1] = \{2\,4\,\overline{5}\,3,\ 4\,\overline{5}\,3\,3,\ \overline{5}\,3\,2\,4,\ 3\,2\,4\,\overline{5},\ \overline{3}\,5\,\overline{4}\,\overline{2},\ \overline{2}\,\overline{3}\,5\,\overline{4},\ \overline{4}\,\overline{2}\,\overline{3}\,5\}$.    □

We say that a signed string $u$ is a **substring** of a circular string $[v]$ if $u$ is a substring of some signed string $w \in [v]$.

## 5.5 Graphs

In the following chapters we shall often use graphs in describing the structure of DNA sequences. We shall consider graphs that allow parallel edges between vertices.

For a finite set $V$, let

$$E(V) = \{\{x, y\} \mid x, y \in V,\ x \neq y\}$$

be the set of all unordered pairs of different elements of $V$.

A **graph** is an ordered triple $\gamma = (V, E, \varepsilon)$, where $V$ and $E$ are finite sets of **vertices**, and **edges**, respectively, and $\varepsilon\colon E \to E(V)$ is the **endpoint mapping** of $\gamma$. If $\varepsilon(e) = \{x, y\}$, then the vertices $x$ and $y$ are the **ends** of $e$.

The set of edges $E$ will be considered as an alphabet in the sense that we consider strings over $E$, and consecutively we use the notation $E^*$.

We draw a graph $\gamma = (V, E, \varepsilon)$ by encircling each vertex $x$ in a circle and presenting each edge $e$ by a line between the ends of $e$.

*Example 5.11.* Let $\gamma = (V, E, \varepsilon)$, where $V = [1, 4]$, $E = \{e_1, e_2, e_3, e_4, e_5\}$ such that $\varepsilon(e_1) = \{1, 2\}$, $\varepsilon(e_2) = \{1, 2\}$, $\varepsilon(e_3) = \{2, 3\}$, $\varepsilon(e_4) = \{2, 4\}$, and $\varepsilon(e_5) = \{3, 4\}$. The graph $\gamma$ is drawn in Fig. 5.1.    □

**Fig. 5.1.** A pictorial representation of the graph of Example 5.11. The vertex set is $[1, 4]$, and there are five edges $E = \{e_1, e_2, e_3, e_4, e_5\}$ for which the endpoints are given by $\varepsilon(e_1) = \{1, 2\}$, $\varepsilon(e_2) = \{1, 2\}$, $\varepsilon(e_3) = \{2, 3\}$, $\varepsilon(e_4) = \{2, 4\}$, and $\varepsilon(e_5) = \{3, 4\}$.

If $e \in E$ is an edge in $\gamma = (V, E, \varepsilon)$ with $\varepsilon(e) = \{x, y\}$, then we say that the vertices $x$ and $y$ are **adjacent**, or that they are **neighbors**. For a vertex $x$ in $\gamma$, let

$$N_\gamma(x) = \{y \mid \varepsilon(e) = \{x, y\},\ e \in E\}$$

be the **neighborhood** of $x$ in $\gamma$. A vertex $x$ is **isolated** in $\gamma$ if $x$ has no neighbors, i.e., if $N_\gamma(x) = \emptyset$.

A graph $\gamma' = (V', E', \varepsilon')$ is a **subgraph** of $\gamma = (V, E, \varepsilon)$ if $\gamma'$ is a part of the graph $\gamma$, i.e., if $V' \subseteq V$, $E' \subseteq E$, and $\varepsilon'(e) = \varepsilon(e)$ for all $e \in E'$. Moreover, a subgraph $\gamma'$ is an **induced subgraph** if every edge $e \in E$ of $\gamma$ that has both ends in $V'$ is also an edge of $\gamma'$.

*Example 5.12.* Let $\gamma = (V, E, \varepsilon)$ be the graph of Example 5.11 (see Fig. 5.1). Let $\gamma' = (V', E', \varepsilon')$ be the graph with $V' = V$, $E' = \{e_1, e_4, e_5\}$, and $\varepsilon'(e_1) = \{1, 2\}$, $\varepsilon'(e_4) = \{2, 4\}$, and $\varepsilon'(e_5) = \{3, 4\}$ (see Fig. 5.2a). Then $\gamma'$ is a subgraph of $\gamma$, but it is not induced since, for instance, $e_2 \notin E'$.

Let then $\gamma'' = (V'', E'', \varepsilon'')$ be the graph with $V'' = \{1, 2, 4\}$, $E'' = \{e_1, e_2, e_4\}$, and $\varepsilon(e_1) = \{1, 2\} = \varepsilon(e_2)$, and $\varepsilon(e_4) = \{2, 4\}$ (see Fig. 5.2b). Then $\gamma''$ is an induced subgraph of $\gamma$. $\qquad\square$

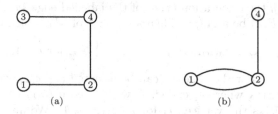

(a)                    (b)

**Fig. 5.2.** The subgraph consisting of the edges $e_1, e_4, e_5$ and their endpoints of the graph $\gamma$ in Fig. 5.1 is given in (a). In (b) the pictured subgraph is an induced subgraph of $\gamma$.

Let $\gamma = (V, E, \varepsilon)$ be a graph, and $K$ a set. A function $f \colon V \to K$ is a **vertex labelling** of $\gamma$, and a function $h \colon E \to K$ is an **edge labelling** of $\gamma$.

Let $f$ be a vertex labelling and $h$ an edge labelling of a graph $\gamma = (V, E, \varepsilon)$. We shall often include the vertex and/or edge labelling in the specification of the graph. Hence we say that $(V, E, \varepsilon, f)$ is a **vertex-labelled graph**, $(V, E, \varepsilon, h)$ is an **edge-labelled graph** and $(V, E, \varepsilon, f, h)$ is a **vertex- and edge-labelled graph**.

*Example 5.13.* Let again $\gamma = (V, E, \varepsilon)$ be as in Example 5.11 (see Fig. 5.1). Let also $f \colon V \to \{a, b\}$ and $h \colon E \to \{0, 1\}$ be the vertex and edge labellings defined by $f(1) = a = f(2)$, $f(3) = b = f(4)$, and $h(e_1) = 1 = h(e_3)$, $h(e_2) = 0 = h(e_4) = h(e_5)$. We have drawn the graph $\gamma$ together with the labelling functions $f$ and $h$ in Fig. 5.3 by putting the label $f(x)$ of a vertex next to the vertex, and by representing each edge with the label 1 by a solid line and each edge with label 0 by a dashed line. $\qquad\square$

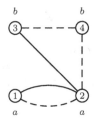

**Fig. 5.3.** The graph $\gamma$ of Example 5.13. The labels $f(x)$ of the vertices are indicated next to the vertices. An edge with label 1 is by represented by a *solid line*, and an edge with label 0 is represented by a *dashed line*.

Let $\gamma = (V, E, \varepsilon, h)$ be an edge-labelled graph, where the labelling is a function $h\colon E \to \{0, 1\}$. Each edge $e \in E$ with $\varepsilon(e) = \{x, y\}$ has two **orientations**: $(x, y)$ and $(y, x)$. If the label $h(e) = i$ (where $i \in \{0, 1\}$), then we shall denote by $x \xrightarrow{i} y$ the orientation $(x, y)$ of the labelled edge $\{x, y\}$.

Let $\gamma = (V, E, \varepsilon)$ be a graph. Then a string $\pi$ of edges such that

$$\pi = e_1 e_2 \ldots e_n \in E^*, \quad \text{with } \varepsilon(e_i) = \{x_i, x_{i+1}\} \in E \text{ for } i = 1, 2, \ldots, n, \quad (5.1)$$

is a **walk**. In a walk $\pi$ the edges $e_i$ can be oriented $(x_i, x_{i+1})$ such that $e_{i+1}$ starts from a vertex where $e_i$ ended. A walk $\pi$ as in (5.1) is a **from** $x_1$ **to** $x_{n+1}$, and it **visits** the vertices $x_i$ for $i \in [1, n+1]$. We also say that the vertices $x_i$ are **on the walk** $\pi$. If no vertex is visited twice in a walk, then it is a **path**. Hence $\pi$ in (5.1) is a path if $x_i \neq x_j$ for all $i \neq j$. Furthermore, if $e_1 e_2 \ldots e_{n-1}$ is a path and $x_{n+1} = x_1$, then $\pi$ is a **cycle**.

Let $h\colon E \to \{0, 1\}$ be an edge labelling of $\gamma$. A path $\pi = e_1 e_2 \ldots e_n$ of $\gamma$ is **alternating** if the labels in the path alternate: $h(e_{2i+1}) = 0$ and $h(e_{2(i+1)}) = 1$ for each $i$. An alternating path is **alternating Hamiltonian**if it visits every vertex of the graph exactly once.

*Example 5.14.* Consider the edge-labelled graph $\gamma$ of Example 5.3. Then $e_1 e_2 e_4 e_3$ is a walk in $\gamma$, but it is not a path since it visits the vertex 2 twice. This walk can be represented as follows:

$$1 \xrightarrow{1} 2 \xrightarrow{1} 3 \xrightarrow{0} 4 \xrightarrow{0} 2.$$

On the other hand, $e_1 e_5$ is not a walk, since $\varepsilon(e_1) \cap \varepsilon(e_5) = \emptyset$. The walk $e_2 e_3 e_4$ is path, and it is alternating, since $h(e_2) = 0$, $h(e_3) = 1$ and $h(e_4) = 0$. In fact, $e_2 e_3 e_4$ is an alternating Hamiltonian path, since it visit every vertex of the graph. $\square$

A graph $\gamma = (V, E, \varepsilon)$ is said to be **simple** if the endpoint mapping $\varepsilon$ is injective, i.e., different edges $e_1, e_2 \in E$ do not share the same ends: $\varepsilon(e_1) \neq \varepsilon(e_2)$. For a simple graph $\gamma$, we shall identify each edge $e \in E$ with its image $\varepsilon(e)$, and thus we can write $\gamma = (V, E)$, where $E \subseteq E(V)$.

*Example 5.15.* The graph in Example 5.11 is not simple, since the edges $e_1$ and $e_2$ have the same ends, 1 and 2. On the other hand, the graph $\gamma = (V, E)$ given in Fig. 5.4 is simple. In this graph the set of vertices is $V = [1, 5]$ and the set of edges is $E = \{\{1, 2\}, \{1, 5\}, \{2, 3\}, \{2, 4\}, \{3, 4\}, \{4, 5\}\}$. □

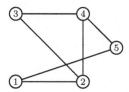

**Fig. 5.4.** The simple graph $\gamma$ of Example 5.15. The edges of $\gamma$ are determined by their endpoints, and thus $\gamma$ can be determined as a pair $(V, E)$ of vertices and pairs between the vertices.

Let $\gamma = (V, E)$ be a simple graph. Then the **complement graph** of $\gamma$ is the graph $\gamma' = (V, E(V) \setminus E)$ obtained from $\gamma$ by interchanging the edges and nonedges.

*Example 5.16.* The complement of the simple graph $\gamma$ of Example 5.15 is given in Fig. 5.5. It has the same vertices as $\gamma$, but its set of edges is $E' = \{\{1, 3\}, \{1, 4\}, \{2, 5\}, \{3, 5\}\}$. □

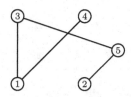

**Fig. 5.5.** The complement of the simple graph $\gamma$ in the previous Example 5.15. Each pair $\{x, y\}$ of different vertices is an edge either in $\gamma$ or its complement.

## Notes on References

- Signed strings are a valuable tool for representing various molecular structures, see, e.g., Pevzner [42].
- For further notions and results in graph theory, we refer to the book of West [58].

# 6

# MDS Arrangements and MDS Descriptors

In this chapter we shall introduce two formalizations of the MDS structure of genes, viz., MDS arrangements and MDS descriptors. MDS arrangements result from representing each elementary or composite MDS by its own symbol, and then by representing the MDS structure of a micronuclear or an intermediate gene by the string of symbols corresponding to the sequence of MDSs. On the other hand, MDS descriptors follow the key observation concerning gene assembly (stated in Chap. 2) that it is not the whole structure of MDSs but only their repeat sequences (pointers) that guide gene assembly. Therefore, in the framework of MDS descriptors each MDS is represented by the pair of its pointers, with each pointer represented by its own symbol. Correspondingly, the MDS descriptor of a micronuclear or an intermediate gene is the string of pairs of pointers corresponding to the sequence of MDSs.

## 6.1 MDS Arrangements

As we have seen in the first part of this book, from the point of view of the process of gene assembly, the structural information about the micronuclear or an intermediate precursor of a macronuclear gene can be given by the sequence of MDSs forming the gene. The whole process can be thought of as a process of assembling MDSs through splicing, to eventually obtain them in the orthodox order $M_1, M_2, \ldots, M_\kappa$. Thus one can represent a macronuclear gene, as well as its micronuclear or an intermediate precursor, by its sequence of MDSs only.

*Example 6.1.* The actin gene of *Sterkiella nova* has the following MDS/IES micronuclear structure (see Fig. 6.1a):

$$M_3 I_1 M_4 I_2 M_6 I_3 M_5 I_4 M_7 I_5 M_9 I_6 \overline{M}_2 I_7 M_1 I_8 M_8. \tag{6.1}$$

By removing the representing IESs, we obtain the "micronuclear pattern": $\alpha = M_3 M_4 M_6 M_5 M_7 M_9 \overline{M}_2 M_1 M_8$, which for our purposes has the same information as the structure given in the representation (6.1).

The macronuclear version of this gene is given in Fig. 6.1b. There the IESs have been excised and the MDSs have been ligated in the orthodox order $M_1, M_2, \ldots, M_9$.                                                                    □

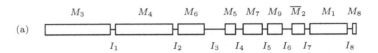

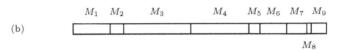

**Fig. 6.1. a** The micronuclear version of the actin gene in *Sterkiella nova.* **b** The macronuclear version of the actin gene of *Sterkiella nova.* The *vertical lines* describe the positions where the MDSs have been spliced together. The final macronuclear gene does not have division into component MDSs $M_1, M_2, \ldots, M_9$.

We use the alphabets

$$\Theta_\kappa = \{\, M_{i,j} \mid 1 \le i \le j \le \kappa\}$$

to denote the MDSs for all $\kappa \ge 1$. Letters $M_{i,i}$ are called **elementary MDSs** and they are also written simply as $M_i$. Each elementary MDS $M_i$ denotes an MDS that is present in the micronucleus. Letters $M_{i,j}$ for $j > i$ are called **composite MDSs**, and they represent those MDSs that are formed during the assembly process (by splicing the MDSs $M_i, M_{i+1}, \ldots, M_j$). Then the sets $\Theta_\kappa^{\maltese}$, for $\kappa \ge 1$, consist of all signed strings representing the sequences of MDSs and their inversions. The elements of $\Theta_\kappa^{\maltese}$ are called **MDS arrangements** (**of size** $\kappa$). We shall denote the MDS arrangements by the letter $\alpha$ with or without indices.

We say that an MDS arrangement $\alpha \in \Theta_\kappa^*$ is **orthodox** if it is of the form

$$\alpha = M_{1,i_2-1} M_{i_2,i_3-1} \ldots M_{i_n,\kappa}. \tag{6.2}$$

The index $\kappa$ is the **size** of $\alpha$. Notice that an orthodox arrangement does not contain any inversions of MDSs. In an orthodox arrangement the (composite) MDSs are in their natural order, and every index $j$ with $1 \le j \le \kappa$ is in exactly one of the intervals $[1, i_2 - 1], [i_2, i_3 - 1], \ldots, [i_{\kappa-1}, i_\kappa - 1], [i_n, \kappa]$.

A signed permutation of an orthodox arrangement $\alpha$ is called a **realistic arrangement**, and it is a **micronuclear arrangement** if it is a signed permutation of the orthodox arrangement $M_1 M_2 \ldots M_\kappa$; that is, if $\alpha = N_1 N_2 \ldots N_\kappa$ such that there is a permutation $(i_1 i_2 \ldots i_\kappa)$ of $[1, \kappa]$ with $N_j = M_{i_j}$ or $N_j = \overline{M}_{i_j}$ for all $j$.

*Example 6.2.* (1) The MDS arrangement $M_{1,2}M_{3,3}\overline{M}_{7,8}M_{4,6}$ over $\Theta_8$ contains an elementary MDS $M_{3,3}$, the composite MDSs $M_{1,2}$, $M_{4,6}$, and the inversion $\overline{M}_{7,8}$ of the composite MDS $M_{7,8}$. Here, for instance, the composite MDS $M_{4,6}$ represents the MDS that is formed from $M_4$, $M_5$, and $M_6$ during the gene assembly process.

(2) The MDS arrangement $M_{6,6}\overline{M}_{4,4}M_{1,1}M_{2,2}\overline{M}_{5,5}M_{3,3}$ contains only elementary MDSs, and it is thus a micronuclear arrangement. It can also be written as $M_6\overline{M}_4M_1M_2\overline{M}_5M_3$.

(3) The MDS arrangement $M_{1,2}M_{3,5}M_{6,7}$ is an orthodox arrangement of size 7. Therefore, for instance, $M_{3,5}\overline{M}_{1,2}M_{6,7}$ is a realistic arrangement.    □

By the following simple result every realistic arrangement is structurally similar to a micronuclear arrangement.

**Lemma 6.1.** *Each realistic arrangement is isomorphic (as a signed string) with a micronuclear arrangement.*

*Proof.* Let $\alpha'$ be a realistic arrangement. Suppose $\alpha'$ is a signed permutation of an orthodox arrangement $\alpha$ as in (6.2). Let $\varphi$ be a morphism of signed strings such that $\varphi(M_{i_r,i_{r+1}-1}) = M_r$ for each $r$ (with the convention that $i_1 = 1$ and $\kappa = i_{n+1} - 1$). Then $\varphi(\alpha) = M_1M_2\ldots M_n$, and therefore $\varphi(\alpha')$ is a signed permutation of $M_1M_2\ldots M_n$. Since $\varphi$ is injective, $\alpha'$ is isomorphic with a micronuclear arrangement, as required by the lemma.    □

## 6.2 MDS Descriptors

Pointers are not present in an assembled gene (because a gene in its macronuclear form does not have any IESs). On the other hand, the micronuclear form of a gene has (possibly many) pointers, and the gene assembly process can be seen as a succession of pointer set removals by molecular operations. Therefore

1. the gene assembly process can be analyzed by representing the micronuclear and each of the intermediate genes by the pattern of pointers present in this gene, and then
2. representing the gene assembly process by the sequence of such patterns, where each next pattern results by the application of molecular operations to the previous one.

Consequently, we can simplify the formal framework by denoting each MDS by the ordered pair of its pointers and markers only.

More formally, let

$$\mathcal{M} = \{\, b, e, \overline{b}, \overline{e} \,\}$$

denote the set of the **markers**. The letter $b$ stands for "beginning," and $e$ stands for "end." For each index $\kappa \geq 2$, we let

$$\Delta_\kappa = \{2, 3, \ldots, \kappa\} \quad \text{and} \quad \Pi_\kappa = \Delta_\kappa \cup \overline{\Delta}_\kappa.$$

Whenever one deals with a specific gene (in any specific species of ciliates) the number of elementary MDSs in the micronucleus determines the index $\kappa$. In order to avoid too involved formalism, we shall assume that, unless explicitly stated otherwise, the index $\kappa$ is clear from the context. Note that, in this context, we do not use the integer 1, because it will represent the begin marker in the following encoding. The elements of the alphabets $\Pi_\kappa$ for $\dot{\kappa} \geq 2$ are called **pointers**. For a pointer $p \in \Pi_\kappa$, the pair $\mathbf{p} = \{p, \overline{p}\}$ is the **pointer set** of $p$.

Now let

$$\Gamma_\kappa = \{(b, e), (b, i), (i, e) \mid 2 \leq i \leq \kappa\} \cup \{(i, j) \mid 2 \leq i < j \leq \kappa\}.$$

We shall consider $\Gamma_\kappa$ as an alphabet, and by convention, for a pair $\delta = (i, j) \in \Gamma_\kappa$, we shall identify its signed version $\overline{\delta}$ with the **inverse pair** $(\overline{j}, \overline{i})$; thus

$$\overline{\Gamma}_\kappa = \{(\overline{e}, \overline{b}), (\overline{i}, \overline{b}), (\overline{e}, \overline{i}) \mid 2 \leq i \leq \kappa\} \cup \{(\overline{j}, \overline{i}) \mid 2 \leq i < j \leq \kappa\}.$$

A signed string $\delta \in \Gamma_\kappa^{\boxplus}$ over $\Gamma_\kappa$ is called an **MDS descriptor**. We shall denote the MDS descriptors by the letter $\delta$ with or without indices and primes. Also, a pointer $p$ **occurs** in $\delta \in \Gamma^{\boxplus}$, if for some pair $\delta'$ in $\delta$ either one of the components is $p$ or its inversion $\overline{p}$.

We define then a morphism $\psi_\kappa \colon \Theta_\kappa^{\boxplus} \to \Gamma_\kappa^{\boxplus}$ as follows:

$$\psi_\kappa(M_{1,\kappa}) = (b, e), \quad \text{and} \quad \psi_\kappa(M_{1,i}) = (b, i+1),$$
$$\psi_\kappa(M_{i,\kappa}) = (i, e), \quad \text{and} \quad \psi_\kappa(M_{i,j}) = (i, j+1), \quad \text{for} \ 1 < i \leq j < \kappa.$$

Note that the morphism $\psi_\kappa$ is a bijection, and therefore for formal purposes $\alpha$ and $\psi_\kappa(\alpha)$ are equivalent for each MDS arrangement $\alpha$ that is, the structure of these two strings is the same. In particular, $\psi_\kappa$ is invertible that is, if $\delta = \psi_\kappa(\alpha)$ then $\alpha = \psi_\kappa^{-1}(\delta)$ is well defined.

*Example 6.3.* The MDS arrangement $\alpha = M_{3,5}\overline{M}_{9,11}\overline{M}_{1,2}M_{12}\overline{M}_{6,8}$ is a signed permutation of the orthodox arrangement $\alpha' = M_{1,2}M_{3,5}M_{6,8}M_{9,11}M_{12}$. By applying the morphism $\psi_{12}$, we obtain the MDS descriptor $\psi_{12}(\alpha) = (3, 6)(\overline{12}, \overline{9})(\overline{3}, \overline{b})(12, e)(\overline{9}, \overline{6})$.  □

We shall now translate terminology for the MDS descriptors from the MDS arrangements. An MDS descriptor $\delta$ is said to be **realistic** if $\delta = \psi_\kappa(\alpha)$ for a realistic arrangement $\alpha$. Moreover, if here $\alpha$ is a micronuclear arrangement, then $\delta$ is a **micronuclear descriptor**. Finally, $\delta$ is an **orthodox MDS descriptor** if $\delta = \psi_\kappa(\alpha)$ for an orthodox arrangement $\alpha$.

Let $\delta_1 = (x_1, x_2) \ldots (x_{2n-1}, x_{2n})$ and $\delta_2 = (y_1, y_2) \ldots (y_{2n-1}, y_{2n})$ be two MDS descriptors in $\Gamma_\kappa^{\maltese}$. They are **isomorphic** if

(1) $x_i \in \mathcal{M} \implies y_i = x_i$,
(2) $x_i \in \Delta_\kappa \iff y_i \in \Delta_\kappa$,
(3) $\|x_i\| < \|x_j\| \iff \|y_i\| < \|y_j\|$, for $x_i, x_j \notin \mathcal{M}$.

If the MDS descriptors $\delta_1$ and $\delta_2$, as in the above, are isomorphic, then the mapping $x_i \mapsto y_i$ is an **isomorphism**.

*Example 6.4.* Let $\delta_1 = (4, 5)(\overline{8}, \overline{6})(b, 4)(8, e)(5, 6)$. Then $\delta_1$ is isomorphic with $\delta_2 = (2, 3)(\overline{5}, \overline{4})(b, 2)(5, e)(3, 4)$. The isomorphism is given by: $4 \mapsto 2$, $5 \mapsto 3$, $8 \mapsto 5$, $6 \mapsto 4$, and $b \mapsto b$, $e \mapsto e$. Note that $\delta_2$ is a micronuclear descriptor, since $\delta_2 = \psi_5(\alpha)$ for the micronuclear arrangement $\alpha = M_2 \overline{M}_4 M_1 M_5 M_3$. $\square$

The following lemma is clear from the definition of the mapping $\psi_\kappa$; see also Lemma 6.1.

**Lemma 6.2.** *An MDS descriptor $\delta$ is realistic if and only if it is isomorphic with a micronuclear descriptor.*

For a realistic MDS descriptor $\delta = (x_1, x_2)(x_3, x_4) \ldots (x_{2n-1}, x_{2n})$ each pointer $p \in \Delta_\kappa$ either does not occur in $\delta$ or it occurs exactly twice in $\delta$. If the latter case holds, then let $\{x_i, x_j\} \subseteq \mathbf{p} = \{p, \overline{p}\}$, for $1 \leq i < j \leq 2n$, be the two occurrences. The $p$-**interval** of $\delta$ is then defined to be the set

$$\delta_{(p)} = \{x_i, x_{i+1}, \ldots, x_j\}.$$

If $x_i = x_j$, then $p$ is **negative** in $\delta$; otherwise (i.e., $x_i = \overline{x}_j$) $p$ is **positive** in $\delta$.

Let $\delta = (p, q)$ be an elements of $\Gamma_\kappa \cup \overline{\Gamma}_\kappa$. If $\delta \in \Gamma_\kappa$, then $p$ is an **incoming pointer** and $q$ is an **outgoing pointer** in $\delta$. On the other hand, if $\delta \in \overline{\Gamma}_\kappa$, then $p$ is an **outgoing pointer** and $q$ is an **incoming pointer** in $\delta$. Note that each pair $\delta = (p, q)$, where $p$ and $q$ are not markers, has one incoming and one outgoing pointer.

We shall now give a characterization of realistic MDS descriptors through intervals of integers. For this purpose, for each $x \in \Pi_\kappa \cup \mathcal{M}$, let

$$\widehat{x} = \begin{cases} 1, & \text{if } x \in \{b, \overline{b}\}, \\ \kappa + 1, & \text{if } x \in \{e, \overline{e}\}, \\ \|x\|, & \text{if } x \in \Pi_\kappa. \end{cases}$$

For each $\delta = (x, y) \in \Gamma_\kappa \cup \overline{\Gamma}_\kappa$, let

$$\widehat{\delta} = [\min\{\widehat{x}, \widehat{y}\}, \max\{\widehat{x}, \widehat{y}\} - 1].$$

Thus $\widehat{\delta}$ is an interval of integers such that $\widehat{\delta} = \widehat{\overline{\delta}}$.

*Example 6.5.* Let $\delta = (4,5)(\overline{8},\overline{6})(b,4)(8,e)(5,6)$. Then the pairs occurring in $\delta$ have the following values: $\widehat{(4,5)} = [4,4]$, $\widehat{(\overline{8},\overline{6})} = [6,7]$, $\widehat{(b,4)} = [1,3]$, $\widehat{(8,e)} = [8,8]$ and $\widehat{(5,6)} = [5,5]$. $\qquad\qquad\qquad\qquad\qquad\qquad\square$

We have then

**Theorem 6.1.** *Let $\delta = \delta_1\delta_2\ldots\delta_n$ be an MDS descriptor, where $\delta_i \in \Gamma_\kappa \cup \overline{\Gamma}_\kappa$ for each $i$. Then $\delta$ is realistic if and only if the intervals $\widehat{\delta}_i$, for $i = 1,2,\ldots,n$, form a partition of the interval $[1,\kappa]$.*

*Proof.* Let $\delta = \psi_\kappa(\alpha)$ for an MDS arrangement $\alpha$. By the definition, $\delta$ is realistic if and only if $\alpha$ is a signed permutation of an orthodox arrangement $\alpha'$. It follows that $\delta$ is realistic if and only if each signed permutation of $\|\delta\|$ is realistic. Hence, without loss in generality, we can assume that $\delta$ has the form $\delta = (x_1,x_2)(x_3,x_4)\ldots(x_{2n-1},x_{2n})$, where, for each $i$, $x_i \in \Delta_\kappa \cup \{b,e\}$ and $\widehat{x}_{2i-1} < \widehat{x}_{2i}$, and, moreover, $\widehat{x}_1 \leq \widehat{x}_3 \leq \cdots \leq \widehat{x}_{2n-1}$. Now, it is clear that $\delta$ is realistic if and only if it is orthodox, that is, if and only if $\alpha$ is an orthodox arrangement. The claim follows now easily from this observation. $\qquad\square$

Let $p,q \in \Pi_\kappa \cup \mathcal{M}$. Then $p$ is **left** and $q$ is **right** in $(p,q)$. For an MDS descriptor $\delta$, $p$ has a **left occurrence (right occurrence)** in $\delta$, if $\delta = \delta_1\delta'\delta_2$, for $\delta_1,\delta_2 \in \Gamma_\kappa^{\maltese}$ and $\delta' \in \Gamma_\kappa \cup \overline{\Gamma}_\kappa$, such that $p$ is left (right) in $\delta'$.

The following properties of realistic MDS descriptors will be useful in the following chapters. The claims of Corollary 6.1 are clear from Theorem 6.1 and the form of the pairs in $\Gamma_\kappa \cup \overline{\Gamma}_\kappa$.

**Corollary 6.1.** *Let $\delta \in \Gamma_\kappa^{\maltese}$ be a realistic MDS descriptor. Then*

*(a) For each marker $b,e \in \mathcal{M}$, $\delta$ has exactly one occurrence from the set $\mathcal{M}$.*
*(b) Each pointer $p \in \Pi_\kappa$ of $\delta$ has exactly two occurrences in $\delta$.*
*(c) For any negative pointer $p \in \Pi_\kappa$ of $\delta$, one occurrence is left and one occurrence is right.*
*(d) For any positive pointer $p \in \Pi_\kappa$, either both occurrences are left or both occurrences are right.*
*(e) No proper scattered substring of $\delta$ is a realistic MDS descriptor.*

Corollary 6.1 yields

**Corollary 6.2.** *For a realistic MDS descriptor $\delta$, if $\delta = \delta_1(x,y)\delta_2$, where $(x,y) = (b,e)$ or $(x,y) = (\overline{e},\overline{b})$ and $\delta_1,\delta_2 \in \Gamma_\kappa^{\maltese}$, then $\delta_1 = \delta_2 = \Lambda$.*

*Proof.* Clearly, $(x,y)$ is a realistic MDS descriptor, which is a scattered substring of $\delta$. By Corollary 6.1(e), $\delta = (x,y)$, i.e., $\delta_1 = \delta_2 = \Lambda$. $\qquad\square$

# Notes on References

- The notions of MDS arrangements and MDS descriptors were introduced by Prescott et al. [22, 51] using somewhat different terminology. Their formal aspects were studied in more detail by Ehrenfeucht et al. [20].
- Lemma 6.1 and Lemma 6.2, which characterize isomorphism of MDS arrangements and MDS descriptors, respectively, are new.
- Corollary 6.1 is a special case of a characterization of realistic MDS descriptors given in [20]. We have adopted in this chapter a more "visual" characterization in Theorem 6.1 using intervals of integers. Also, Corollary 6.2 is due to [20].

# 7

# MDS Descriptor Pointer Reduction System

In Chap. 6 we formalized the MDS structure of micronuclear and intermediate genes through MDS arrangements and MDS descriptors. In this chapter we shall continue to consider the framework of MDS descriptors. In particular, we shall formalize in this framework the use of the three molecular operations $ld$, $hi$, and $dlad$ for gene assembly introduced in Chap. 3. Thus, these operations will be formalized as operations on (rewriting rules for) MDS descriptors. We also prove that this formal model is universal, i.e., every (realistic) MDS descriptor can be reduced to the MDS descriptor $(b, e)$ or its inverse $(\overline{e}, \overline{b})$ by these three operations.

## 7.1 Assembly Operations on MDS Descriptors

In the following we define three types of operations ld, hi, and dlad on MDS descriptors. Note that the MDS descriptors involved in the definitions need not be realistic. In this sense, the formal operations for gene assembly apply to somewhat more general cases than the ones required by the molecular operations of Chap. 3.

There are altogether 11 cases for the operations (two for both ld and hi, and seven for dlad). The different cases arise from different possibilities how the MDSs (represented by elements of the sets $\Gamma_\kappa$) are positioned with respect to each other in the MDS descriptor. These possibilities are portrayed in Figs. 7.1–7.5. In these illustrations

- a rectangle denotes an element of $\Gamma_\kappa$ (i.e., an MDS with its pointers indicated)
- a zigzag line denotes an arbitrary string in $\Gamma_\kappa^\circledast$ (i.e., a segment of a molecule that may contain both MDSs and IESs), and
- a straight line segment denotes adjacency of elements in $\Gamma_\kappa$ (i.e., one IES)

(1) For each pointer $p \in \Pi_\kappa$, the ld-**rule** for $p$ is defined as follows:

$$\mathsf{ld}_p(\delta_1(q,p)(p,r)\delta_2) = \delta_1(q,r)\delta_2, \tag{$\ell 1$}$$

$$\mathsf{ld}_p((p,q)\delta_1(r,p)) = (r,q)\delta_1, \tag{$\ell 2$}$$

where $q, r \in \Pi_\kappa \cup \mathcal{M}$, and $\delta_1, \delta_2 \in \Gamma_\kappa^{\maltese}$.

Note that pointer $p$ must be negative in $\delta$ in order for $\mathsf{ld}_p$ to be applicable to the MDS descriptor $\delta$. Let

$$\mathsf{Ld} = \{\mathsf{ld}_p \mid p \in \Pi_\kappa, \ \kappa \geq 2\}$$

be the set of all ld-rules.

Case ($\ell 1$) is called a **simple ld-rule**. This case corresponds to the molecular case, where two adjacent occurrences of pointer $p$ are separated by one IES only. The simple ld-rule is illustrated in Fig. 7.1.

**Fig. 7.1.** The MDS/IES structure to which simple $\mathsf{ld}_p$ is applicable. The *rectangles* represent MDSs with pointers indicated. A *zigzag line* represents a segment of the molecule that may contain both MDSs and IESs. A *straight line* segment represents one IES.

Case ($\ell 2$) is called a **boundary ld-rule**. The boundary case represents the molecular case, where pointer $p$ occurs at the boundaries of the part of the molecule that contains all MDSs. (Note that there may be some IESs present outside the boundary.) The boundary ld-rule is illustrated in Fig. 7.2. Later, see p. 79, we show that we can restrict to the special case of the boundary ld-rule, where the middle portion $\delta_1$ consists of one IES only.

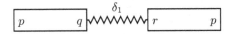

**Fig. 7.2.** The MDS/IES structure to which boundary $\mathsf{ld}_p$ is applicable. The boundary $\mathsf{ld}_p$-rule represents the case, where the two occurrences of pointer $p$ are at the boundaries of the part of the molecule that contains all MDSs.

(2) For each pointer $p \in \Pi_\kappa$, the hi-**rule** for $p$ is defined as follows:

$$\mathsf{hi}_p(\delta_1(p,q)\delta_2(\bar{p},\bar{r})\delta_3) = \delta_1\bar{\delta_2}(\bar{q},\bar{r})\delta_3, \tag{h1}$$

$$\mathsf{hi}_p(\delta_1(q,p)\delta_2(\bar{r},\bar{p})\delta_3) = \delta_1(q,r)\bar{\delta_2}\delta_3, \tag{h2}$$

where $q, r \in \Pi_\kappa \cup \mathcal{M}$, and $\delta_i \in \Gamma_\kappa^{\maltese}$, for each $i = 1, 2, 3$.

Note that here the pointer $p$ must be positive in the MDS descriptor to which $\mathrm{hi}_p$ is applicable. The $\mathrm{hi}_p$-rule is illustrated in Fig. 7.3. Let

$$\mathsf{Hi} = \{\mathrm{hi}_p \mid p \in \Pi_\kappa,\ \kappa \geq 2\}$$

be the set of all hi-rules.

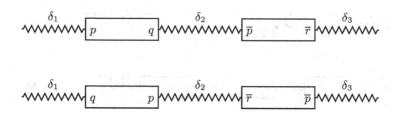

**Fig. 7.3.** The MDS/IES structures to which the rule $\mathrm{hi}_p$ is applicable. The pointer $p$ is positive, and the occurrences of $p$ are either left or they are both right.

(3) For a pair $p, q \in \Pi_\kappa$ of different pointers, the **dlad-rule** for $p$ and $q$ is defined as follows:

$$\mathrm{dlad}_{p,q}(\delta_1(p, r_1)\delta_2(q, r_2)\delta_3(r_3, p)\delta_4(r_4, q)\delta_5) = \delta_1\delta_4(r_4, r_2)\delta_3(r_3, r_1)\delta_2\delta_5, \quad (\mathrm{d}1)$$

$$\mathrm{dlad}_{p,q}(\delta_1(p, r_1)\delta_2(r_2, q)\delta_3(r_3, p)\delta_4(q, r_4)\delta_5) = \delta_1\delta_4\delta_3(r_3, r_1)\delta_2(r_2, r_4)\delta_5, \quad (\mathrm{d}2)$$

$$\mathrm{dlad}_{p,q}(\delta_1(r_1, p)\delta_2(q, r_2)\delta_3(p, r_3)\delta_4(r_4, q)\delta_5) = \delta_1(r_1, r_3)\delta_4(r_4, r_2)\delta_3\delta_2\delta_5, \quad (\mathrm{d}3)$$

$$\mathrm{dlad}_{p,q}(\delta_1(r_1, p)\delta_2(r_2, q)\delta_3(p, r_3)\delta_4(q, r_4)\delta_5) = \delta_1(r_1, r_3)\delta_4\delta_3\delta_2(r_2, r_4)\delta_5, \quad (\mathrm{d}4)$$

$$\mathrm{dlad}_{p,q}(\delta_1(p, r_1)\delta_2(q, p)\delta_4(r_4, q)\delta_5) = \delta_1\delta_4(r_4, r_1)\delta_2\delta_5, \quad (\mathrm{d}5)$$

$$\mathrm{dlad}_{p,q}(\delta_1(p, q)\delta_3(r_3, p)\delta_4(q, r_4)\delta_5) = \delta_1\delta_4\delta_3(r_3, r_4)\delta_5, \quad (\mathrm{d}6)$$

$$\mathrm{dlad}_{p,q}(\delta_1(r_1, p)\delta_2(q, r_2)\delta_3(p, q)\delta_5) = \delta_1(r_1, r_2)\delta_3\delta_2\delta_5, \quad (\mathrm{d}7)$$

where $r_i \in \Pi_\kappa \cup \mathcal{M}$ and $\delta_i \in \Gamma_\kappa^{\maltese}$ for each $i$.

In each of the above instances of $\mathrm{dlad}_{p,q}$, pointers $p$ and $q$ overlap, and they both must be negative in the MDS descriptor to which $\mathrm{dlad}_{p,q}$ is applicable. Let

$$\mathsf{Dlad} = \{\mathrm{dlad}_{p,q} \mid p, q \in \Pi_\kappa,\ \kappa \geq 2\}$$

be the set of all dlad-rules.

In the above definition of dlad, we distinguish first Cases (d1)–(d4), illustrated in Fig. 7.4. In these cases all four occurrences of $p$ and $q$ are in different MDSs. We also have the following "short" cases (d5), (d6), and (d7), illustrated in Fig. 7.5. In these cases, one occurrence of $p$ and one of $q$ are in the

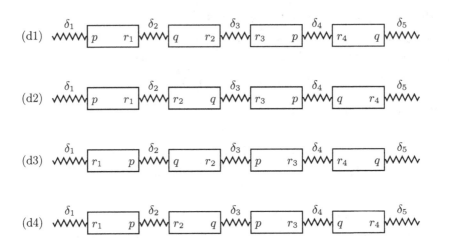

**Fig. 7.4.** The MDS/IES structures (d1)–(d4) to which $\mathsf{dlad}_{p,q}$ is applicable. These cases involve four disjoint elements of $\Gamma_\kappa$ (representing MDSs). The pointers $p$ and $q$ overlap with each other, and they are both negative.

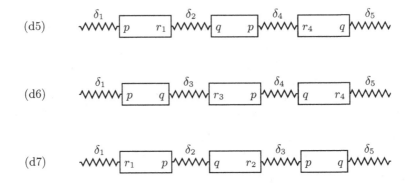

**Fig. 7.5.** The MDS/IES structures (d5), (d6), and (d7) to which $\mathsf{dlad}_{p,q}$ is applicable. These cases involve three disjoint elements of $\Gamma_\kappa$ (representing MDSs). The pointers $p$ and $q$ overlap with each other, and they are both negative.

same MDS. We note that the two occurrences of $p$ (or $q$ ) can never be in the same MDS.

The following basic result will be often used implicitly in the following material.

**Lemma 7.1.** *Let $\delta$ be a realistic MDS descriptor, let $p$ and $q$ be pointers in $\delta$, and let $\varphi \in \mathsf{Ld} \cup \mathsf{Hi} \cup \mathsf{Dlad}$. If $\varphi$ is applicable to $\delta$, then $\varphi(\delta)$ is also realistic.*

*Proof.* The claim is clear by the form of the operations and Theorem 6.1, which characterizes the realistic MDS descriptors.                                    □

For an MDS descriptor $\delta$ and operations $\varphi_1, \ldots, \varphi_n$ with $n \geq 1$ from $\mathsf{Ld} \cup \mathsf{Hi} \cup \mathsf{Dlad}$, a composition $\varphi = \varphi_\kappa \ldots \varphi_1$ is an **assembly strategy** for $\delta$, if $\varphi$ is applicable to $\delta$ (i.e., $\varphi(\delta)$ is well defined). Also, $\varphi$ is **successful** for $\delta$ if either $\varphi(\delta) = (b, e)$ or $\varphi(\delta) = (\overline{e}, \overline{b})$.

*Example 7.1.* (1) The MDS descriptor $\delta = (4, 5)(\overline{2}, \overline{b})(5, e)(\overline{4}, \overline{3})(\overline{3}, \overline{2})$ is realistic. Indeed, $\delta = \psi_5(\alpha)$ for the micronuclear arrangement $\alpha = M_4 \overline{M}_1 M_5 \overline{M}_3 \overline{M}_2$. Each of the operations $\mathsf{Id}_{\overline{3}}$, $\mathsf{hi}_4$, and $\mathsf{dlad}_{5,\overline{2}}$ is applicable to $\delta$, and the results of these applications are

$$\mathsf{Id}_{\overline{3}}(\delta) = (4, 5)(\overline{2}, \overline{b})(5, e)(\overline{4}, \overline{2}),$$
$$\mathsf{hi}_4(\delta) = (\overline{e}, \overline{5})(b, 2)(\overline{5}, \overline{3})(\overline{3}, \overline{2}),$$
$$\mathsf{dlad}_{5,\overline{2}}(\delta) = (4, e)(\overline{4}, \overline{3})(\overline{3}, \overline{b}).$$

Also, $\mathsf{hi}_4 \, \mathsf{dlad}_{5,\overline{2}}(\delta) = (\overline{e}, \overline{3})(\overline{3}, \overline{b})$ and $\mathsf{Id}_{\overline{3}} \, \mathsf{hi}_4 \, \mathsf{dlad}_{5,\overline{2}}(\delta) = (\overline{e}, \overline{b})$, and hence $\mathsf{Id}_{\overline{3}} \, \mathsf{hi}_4 \, \mathsf{dlad}_{5,\overline{2}}$ is a successful assembly strategy for $\delta$.

(2) Consider the realistic MDS descriptor $\delta = (5, e)(b, 2)(2, 3)(\overline{5}, \overline{4})(3, 4)$. The $\mathsf{Id}_2$-rule is applicable to $\delta$, and $\mathsf{Id}_2(\delta) = (5, e)(b, 3)(\overline{5}, \overline{4})(3, 4)$. Also, the operation $\mathsf{hi}_5$ is applicable to $\delta$, and $\mathsf{hi}_5(\delta) = (\overline{3}, \overline{2})(\overline{2}, \overline{b})(\overline{e}, \overline{4})(3, 4)$.

The composition $\mathsf{Id}_5 \, \mathsf{hi}_3 \, \mathsf{hi}_{\overline{4}} \, \mathsf{Id}_2$ yields the following sequence of realistic MDS descriptors:

$$\mathsf{Id}_2(\delta) = (5, e)(b, 3)(\overline{5}, \overline{4})(3, 4),$$
$$\mathsf{hi}_{\overline{4}} \, \mathsf{Id}_2(\delta) = (5, e)(b, 3)(\overline{5}, \overline{3}),$$
$$\mathsf{hi}_3 \, \mathsf{hi}_{\overline{4}} \, \mathsf{Id}_2(\delta) = (5, e)(b, 5),$$
$$\mathsf{Id}_5 \, \mathsf{hi}_3 \, \mathsf{hi}_{\overline{4}} \, \mathsf{Id}_2(\delta) = (b, e),$$

and thus it is successful for $\delta$. Notice that we did not use any dlad-rules in this assembly strategy.                                                           □

We come back now to the boundary Id-rule. Clearly, if we have a boundary occurrence of pointer $p$ in an MDS descriptor $\delta = (p, q)\delta_1(r, p)$ as in ($\ell 2$), then the only rule applicable to $p$ is the boundary Id-rule. Moreover, if a rule $\varphi$ is applied to any other pointer or pointers (in $\delta_1$), then $\varphi(\delta)$ still has pointer $p$ as its boundary. Therefore, as far as assembly of the macronuclear gene is concerned, we may assume that the Id-rule for the boundary pointer $p$ is applied as the last one in the assembly. Hence the situation just before the boundary Id-rule is given in ($\ell 2'$), and it is illustrated in Fig. 7.6:

$$\mathsf{Id}_p((p, m)(m', p)) = (m', m), \qquad (\ell 2')$$

where $m, m' \in \mathcal{M}$. Consequently, in Part II we shall assume that *the boundary $\mathsf{Id}_p$-rule for pointer $p$ will have the form* ($\ell 2'$).

The reader is referred to Chap. 12 in Part III where the topic of circular assembly using boundary rules is treated in depth.

**Fig. 7.6.** The MDS/IES structure to which the boundary rule $(\ell2')$ is applicable. The first MDS ends with marker $m$ and the second MDS begins with marker $m'$.


## 7.2 The Assembling Power of the Operations

Each application of an operation from Ld $\cup$ Hi $\cup$ Dlad shortens the involved MDS descriptor. A single use of a ld-rule and a hi-rule shortens the MDS descriptor by one pair, and a single use of a dlad-rule shortens the descriptor by two pairs. Recall also that a successful assembly leads to a macronuclear MDS descriptor $(b, e)$ or $(\bar{e}, \bar{b})$, which has no pointers.

We prove now that the set of our three operations on MDS descriptors, ld, hi, and dlad, is *universal*, i.e., any realistic MDS descriptor $\delta$ has a successful assembly strategy for it.

**Theorem 7.1.** *Each realistic MDS descriptor has a successful assembly strategy.*

*Proof.* We prove first that for every realistic MDS descriptor $\delta$ (different from $(b, e)$ and $(\bar{e}, \bar{b})$) at least one operation from Ld $\cup$ Hi $\cup$ Dlad is applicable to $\delta$. Let then $\delta$ be such an MDS descriptor, and assume that no operation from Ld $\cup$ Hi is applicable to $\delta$.

Since no rule from Hi is applicable, all pointers in $\delta$ must be negative, and since no rule from Ld is applicable, $\delta$ has no simple direct repeat pattern $(q, p)(p, r)$ nor $(p, m)(m', p)$ for markers $m$ and $m'$. (Recall that for the boundary case we apply $(\ell2')$ from p. 79.) Consequently, if pointer $p$ occurs in $\delta$, then the $p$-interval $\delta_{(p)}$ must contain at least one other pointer.

Let then $p$ be a pointer in $\delta$ such that the number of pointers within the $p$-interval is minimum, that is, no pointer from $\delta$ has less pointers in its interval. Let $q$ be a pointer that has an occurrence within the $p$-interval. Now, $\delta$ contains two occurrences of $q$, and the other occurrence of $q$ must be outside the $p$-interval, as otherwise the minimality assumption on $p$ is contradicted. Moreover, since all pointers in $\delta$ are negative, either $\text{dlad}_{p,q}$ or $\text{dlad}_{q,p}$ is applicable to $\delta$.

Since applying any of our operations decreases the number of pointers, and, by Lemma 7.1, applying any of them to a realistic MDS descriptor yields again a realistic MDS descriptor, the theorem holds. $\qquad\square$

To ensure the existence of a successful assembly strategy for *all* realistic MDS descriptors, one needs all three types of operations. Thus, e.g., the realistic MDS descriptor $(b, 2)(2, e)$ needs the operation $\text{ld}_2$ from Ld, $(\bar{2}, \bar{b})(2, e)$ needs the operation $\text{hi}_{\bar{2}}$ from Hi, and $(b, 2)(3, e)(2, 3)$ needs the operation $\text{dlad}_{2,3}$ from Dlad.

## Notes on References

• The formal operations ld, hi, and dlad on the MDS descriptors were introduced by Ehrenfeucht et al. [19], where also the universality result, Theorem 7.1, was proved. The rather elementary result of Lemma 7.1 is new.

# 8

## Legal Strings

We move now to consider a substantial simplification of representing micronuclear and intermediate genes. We shall use only the sequence of pointers in the order in which they appear in such a gene, obtaining in this way just a string of pointers. This yields the framework of legal strings, which is more abstract than the framework of MDS descriptors because different MDS descriptors may yield the same legal string. In this chapter we investigate the basic properties of the representation of the MDS structure by legal strings. In particular, we characterize those legals string that either correspond directly to *realistic* MDS descriptors or that can be signed so as to correspond to realistic MDS descriptors.

## 8.1 Representation by Legal Strings

A string $v \in \Sigma^*$ over an (unsigned) alphabet $\Sigma$ is a **double occurrence string** if every letter $a \in \mathrm{dom}(v)$ occurs exactly twice in $v$. A signing of a nonempty double occurrence string is a **legal string**.

Let $a \in \Sigma \cup \overline{\Sigma}$ and let $u \in \Sigma^{\circledast}$ be a legal string. If $u$ contains both substrings $a$ and $\overline{a}$, then $a$ is **positive** in $u$; otherwise, $a$ is **negative** in $u$.

*Example 8.1.* Consider the signed string $u = 2\,4\,3\,\overline{2}\,\overline{5}\,3\,4\,5$ over $\Delta_5$. Clearly, $u$ is legal. Pointers 2 and 5 are positive in $u$, while 3 and 4 are negative in $u$. On the other hand, the string $w = 2\,4\,3\,\overline{2}\,\overline{5}\,3\,5$ is not legal, since 4 has only one occurrence in $w$. □

Let $u = a_1 a_2 \ldots a_n \in \Sigma^{\circledast}$ be a legal string over $\Sigma$, where $a_i \in \Sigma \cup \overline{\Sigma}$ for each $i$. For each letter $a \in \mathrm{dom}(u)$, there are indices $i$ and $j$ with $1 \leq i < j \leq n$ such that $\|a_i\| = a = \|a_j\|$. The substring

$$u_{(a)} = a_i a_{i+1} \ldots a_j$$

is the $a$-**interval** of $u$. Two different letters $a, b \in \Sigma$ are said to **overlap in** $u$ if the $a$-interval and the $b$-interval of $u$ overlap: if $u_{(a)} = a_{i_1} \ldots a_{j_1}$ and

$u_{(b)} = a_{i_2} \ldots a_{j_2}$, then either $i_1 < i_2 < j_1 < j_2$ or $i_2 < i_1 < j_2 < j_1$. Moreover, for each letter $a$, we denote by

$$O_u(a) = \{b \in \Sigma \mid b \text{ overlaps with } a \text{ in } u\} \cup \{a\}.$$

Similarly, denote by $O_u^+(a)$ ($O_u^-(a)$, resp.) the set of all positive letters (negative letters, resp.) in $O_u(a)$. For technical reasons, it is convenient to include $a$ in $O_u(a)$. Hence if $a$ is positive in $u$, then $a \in O_u^+(a)$, and if $a$ is negative in $u$, then $a \in O_u^-(a)$.

*Example 8.2.* Let $u = 2\,4\,3\,5\,3\,\overline{2}\,\overline{6}\,\overline{5}\,7\,4\,6\,7$ be a string of pointers. The 2-interval of $u$ is the substring $u_{(2)} = 2\,4\,3\,5\,3\,\overline{2}$, which contains only one occurrence of pointer 4 and pointer 5, but either two or no occurrences of $3, 6$, and 7. Hence pointer 2 overlaps with 4 and 5 but not with $3, 6$ and 7, and so $O_u(2) = \{2, 4, 5\}$, $O_u^+(2) = \{2, 5\}$, and $O_u^-(2) = \{4\}$. Similarly,

$$
\begin{aligned}
u_{(3)} &= 3\,5\,3, & &\text{and } 3 \text{ overlaps with } 5, \\
u_{(4)} &= 4\,3\,5\,3\,\overline{2}\,\overline{6}\,\overline{5}\,7\,4, & &\text{and } 4 \text{ overlaps with } 2, 6, 7, \\
u_{(5)} &= 5\,3\,\overline{2}\,\overline{6}\,\overline{5}, & &\text{and } 5 \text{ overlaps with } 2, 3, 6, \\
u_{(6)} &= \overline{6}\,\overline{5}\,7\,4\,6, & &\text{and } 6 \text{ overlaps with } 4, 5, 7, \\
u_{(7)} &= 7\,4\,6\,7, & &\text{and } 7 \text{ overlaps with } 4, 6.
\end{aligned}
$$

□

We shall now simplify the MDS descriptors by removing the parentheses from the descriptors and deleting the markers. We obtain in this way legal strings as the formalism for describing the sequences of pointers present in the micronuclear and the intermediate molecules.

Let $\mathrm{rem}_\kappa \colon \Gamma_\kappa^{\maltese} \to \Delta_\kappa^{\maltese}$ be the morphism that removes the parentheses and the markers from the string of pairs. More formally, let

$$
\mathrm{rem}_\kappa(x, y) = \begin{cases}
\Lambda, & \text{if } x, y \in \mathcal{M}, \\
x, & \text{if } x \in \Delta_\kappa \cup \overline{\Delta}_\kappa, \ y \in \mathcal{M}, \\
y, & \text{if } x \in \mathcal{M}, \ y \in \Delta_\kappa \cup \overline{\Delta}_\kappa, \\
xy, & \text{if } x, y \in \Delta_\kappa \cup \overline{\Delta}_\kappa,
\end{cases}
$$

for all $(x, y) \in \Gamma_\kappa \cup \overline{\Gamma}_\kappa$.

*Example 8.3.* Let $\delta = (4, 5)(\overline{2}, \overline{b})(5, e)(\overline{4}, \overline{2})$ be an MDS descriptor. Then we have that $\mathrm{rem}_5(4, 5) = 4\,5$, $\mathrm{rem}_5(\overline{2}, \overline{b}) = \overline{2}$, $\mathrm{rem}_5(5, e) = 5$, and $\mathrm{rem}_5(\overline{4}, \overline{2}) = \overline{4}\,\overline{2}$. Therefore $\mathrm{rem}_5(\delta) = 4\,5\,\overline{2}\,5\,\overline{4}\,\overline{2}$. □

For MDS arrangements, we define a morphism $\varrho_\kappa : \Theta_\kappa^{\maltese} \to \Delta_\kappa^{\maltese}$ by letting, for each $\alpha \in \Theta_\kappa^{\maltese}$,

$$\varrho_\kappa(\alpha) = \mathrm{rem}_\kappa(\psi_\kappa(\alpha)).$$

Here each MDS arrangement $\alpha \in \Theta_\kappa^{\maltese}$ is first mapped onto the MDS descriptor $\psi_\kappa(\alpha)$, which is then mapped onto the legal string $\mathrm{rem}_\kappa(\psi_\kappa(\alpha))$. In particular, for the elementary MDSs, we have

$$\varrho_\kappa(M_1) = 2, \quad \varrho_\kappa(M_\kappa) = \kappa, \quad \varrho_\kappa(M_i) = i\,i{+}1, \quad \text{for } 2 < i < \kappa,$$

and $\varrho_\kappa(\overline{M}_i) = \overline{\varrho_\kappa(M_i)}$ for $1 \leq i \leq \kappa$.

We say that a legal string $u$ is **realistic** if there exists a realistic MDS descriptor $\delta$ such that $u = \mathrm{rem}_\kappa(\delta)$ for some $\kappa$, or equivalently $u = \varrho_\kappa(\alpha)$ for a realistic arrangement $\alpha$.

*Example 8.4.* Consider the micronuclear arrangement of the actin gene of *Sterkiella nova*: $\alpha = M_3 M_4 M_6 M_5 M_7 M_9 \overline{M}_2 M_1 M_8$. We have first that

$$\psi_9(\alpha) = (3,4)(4,5)(6,7)(5,6)(7,8)(9,e)(\overline{3},\overline{2})(b,2)(8,9),$$

and then

$$\varrho_9(\alpha) = 34\ 45\ 67\ 56\ 78\ 9\ \overline{3}\,\overline{2}\ 2\ 89 = \mathrm{rem}_9\,\psi_9(\alpha).$$

Since $\alpha$ is a realistic arrangement, $\psi_9(\alpha)$ is a realistic MDS descriptor, and, consequently, $\varrho_9(\alpha) = \mathrm{rem}_9\,\psi_9(\alpha)$ is a realistic legal string.  □

The following example shows that there exist legal strings that are not realistic.

*Example 8.5.* (1) The string $u = 2\,3\,4\,3\,2\,4$ is legal, but it is not realistic, since it has no "realistic parsing". Indeed, suppose that there exists a realistic arrangement $\alpha$ such that $u = \varrho_4(\alpha)$. Since $\mathrm{dom}(u) = [2,4]$, $\alpha$ must be a micronuclear arrangement. Then $\alpha$ must end with the MDS $M_4$, since $\varrho_4(M) \neq 24$ for all MDSs $M$. Similarly, since $\varrho_4(M) \neq 32$ for all MDSs $M$, $\alpha$ must end with $M_1 M_4$. Now, however, $u = 2\,3\,4\,3\,\varrho_4(M_1 M_4)$ gives a contradiction, since 3 and 43 are not images of any MDSs.

(2) The legal string $u = 22$ is realistic as it corresponds to the orthodox arrangement $M_1 M_2$ that is, $\varrho_2(M_1 M_2) = \mathrm{rem}_2((b,2)(2,e)) = 22$.  □

## 8.2 Realizable Legal Strings

Although the essential information about the gene assembly process is preserved in the abstraction processes, some structural information about the molecule is lost. We characterize in the following those pairs of micronuclear arrangements that have the same realistic legal string describing them.

Consider micronuclear arrangements of size $\kappa$, and let

$$(k_1, k_2) = \begin{cases} (\kappa, \kappa - 1), & \text{if } \kappa \text{ is odd,} \\ (\kappa - 1, \kappa), & \text{if } \kappa \text{ is even.} \end{cases}$$

Also, let

$$\alpha_o = M_1 M_3 \ldots M_{k_1}, \quad \text{and} \quad \alpha_e = M_2 M_4 \ldots M_{k_2}. \tag{8.1}$$

Hence the orthodox arrangement $M_1 M_2 \ldots M_\kappa$ is obtained by performing a perfect shuffle of the two sequences $\alpha_o$ and $\alpha_e$.

**Lemma 8.1.** *Let $\alpha_1$ and $\alpha_2$ be two different MDS arrangements of the same size $\kappa$ such that $\varrho_\kappa(\alpha_1') \neq \varrho_\kappa(\alpha_2')$ for all proper prefixes $\alpha_1'$ of $\alpha_1$ and $\alpha_2'$ of $\alpha_2$. If $\varrho_\kappa(\alpha_1) = \varrho_\kappa(\alpha_2)$, then $\{\alpha_1, \alpha_2\} = \{\alpha_o, \alpha_e\}$, or $\{\alpha_1, \alpha_2\} = \{\overline{\alpha}_o, \overline{\alpha}_e\}$.*

*Proof.* By the hypothesis, $\alpha_1$ and $\alpha_2$ begin with different MDSs, say $\alpha_1 = N_1 \alpha_{11}$ and $\alpha_2 = N_2 \alpha_{21}$, where $N_1 \neq N_2$. Now $\varrho_\kappa(\alpha_1) = \varrho_\kappa(\alpha_2)$ and the condition on the prefixes implies that either $\varrho_\kappa(N_1)$ or $\varrho_\kappa(N_2)$ is a proper prefix of the other. By symmetry, we can assume that $\varrho_\kappa(N_1)$ is a prefix of $\varrho_\kappa(N_2)$. By the definition of the mapping $\varrho_\kappa$, we have that either

(i) $N_1 = M_1$, and $N_2 = M_2$, so that $\varrho_\kappa(N_1) = 2$ and $\varrho_\kappa(N_2) = 23$, or
(ii) $N_1 = \overline{M}_\kappa$, and $N_2 = (\overline{M}_{\kappa-1})$, so that $\varrho_\kappa(N_1) = \overline{\kappa}$ and $\varrho_\kappa(N_2) = \overline{\kappa}\,(\overline{\kappa - 1})$.

In Case (i), $\varrho_\kappa(\alpha_1) = 2\varrho_\kappa(\alpha_{11}) = 23\varrho_\kappa(\alpha_{21}) = \varrho_\kappa(\alpha_2)$, and thus necessarily $\alpha_{11}$ begins with $M_3$ (since $\varrho_\kappa(M_3) = 34$), and then $\alpha_{21}$ begins with $M_4$. It follows inductively that $\alpha_1 = M_1 \alpha_{11} = \alpha_o$ and $\alpha_2 = M_2 \alpha_{21} = \alpha_e$.

Case (ii) reduces to Case (i), since now the inverses $\overline{v}_1$ and $\overline{v}_2$ necessarily satisfy the condition (i).    □

The following result is a simple consequence of Lemma 8.1.

**Theorem 8.1.** *Let $\alpha_1$ and $\alpha_2$ be different micronuclear arrangements of the same size $\kappa$. Then $\varrho_\kappa(\alpha_1) = \varrho_\kappa(\alpha_2)$ if and only if one of the following four conditions holds:*

$$\alpha_1 = \alpha_o \alpha_e, \quad \text{and} \quad \alpha_2 = \alpha_e \alpha_o, \tag{i}$$

$$\alpha_1 = \alpha_o \overline{\alpha}_e, \quad \text{and} \quad \alpha_2 = \alpha_e \overline{\alpha}_o, \tag{ii}$$

$$\alpha_1 = \overline{\alpha}_o \alpha_e, \quad \text{and} \quad \alpha_2 = \overline{\alpha}_e \alpha_o, \tag{iii}$$

$$\alpha_1 = \overline{\alpha}_o \overline{\alpha}_e, \quad \text{and} \quad \alpha_2 = \overline{\alpha}_e \overline{\alpha}_o. \tag{iv}$$

*Example 8.6.* For $\alpha_1 = M_1 M_3 M_5 \overline{M}_4 \overline{M}_2$ and $\alpha_2 = M_2 M_4 \overline{M}_5 \overline{M}_3 \overline{M}_1$, we have, indeed, that $\varrho_5(\alpha_1) = \varrho_5(\alpha_2)$, because $\alpha_1 = \alpha_o \overline{\alpha}_e$ and $\alpha_2 = \alpha_e \overline{\alpha}_o$, where $\alpha_e = M_2 M_4$ and $\alpha_o = M_1 M_3 M_5$.    □

A legal string $u$ over an alphabet $\Sigma$ is said to be **realizable** if $u$ is isomorphic to a realistic legal string, that is, if there exists an injective morphism $\tau \colon \Sigma^\maltese \to \Delta_\kappa^\maltese$, for some $\kappa \geq 2$, such that $\tau(\Sigma) \subseteq \Delta_\kappa$ and $\tau(u)$ is realistic.

*Example 8.7.* (1) The signed string $u = 5\,4\,\overline{3}\,2\,\overline{5}\,\overline{2}\,4\,3$ with $\mathrm{dom}(u) = \Sigma = [2,5]$ is legal. It is not realistic, since, obviously, $4\,\overline{3}\,2$ can never be a substring of a realistic legal string. However, $u$ is realizable: if the morphism $\tau$ is defined by $\tau(2) = 2$, $\tau(3) = 5$, $\tau(4) = 4$, and $\tau(5) = 3$, then $\tau(u) = 3\,4\,\overline{5}\,2\,\overline{3}\,\overline{2}\,4\,5$, and it is realistic: $\tau(u) = \varrho_5(\alpha)$ for the realistic arrangement $\alpha = M_3\overline{M}_5M_1\overline{M}_2M_4$ (which is even a micronuclear arrangement).

(2) The signed string $u = 2\,2\,3\,4\,4\,3\,5\,5$ is not realizable. Indeed, suppose there exists a realistic legal string $v$ and a morphism $\tau$ such that $u = \tau(v)$. By Lemma 6.1, each realistic arrangement is isomorphic with a micronuclear arrangement, and hence we may assume that $\mathrm{dom}(v) = [2,5]$. Since a string $pp$ is never an image $\varrho_5(M_i)$ of any MDS $M_i$, either (2i) or (2ii) holds:

(2i) $\tau(2) = 2$ and $\tau(5) = 5$. Now also $\tau(3) = 3$, since $u$ begins with $2(23)$. This yields a contradiction, since now $35$ is a substring of $v$.

(2ii) $\tau(2) = 5$ and $\tau(5) = 2$. Now, $v$ should begin with $55$, but $55$ is never a prefix of any realistic legal string (although $\overline{5}\,\overline{5}$ can be).

(3) A simple analysis shows that the string $u = 2\,3\,2\,4\,\overline{3}\,4$ is not realizable. Note that $u$ has no legal substrings and no conjugate of $u$ is realizable. □

We now relax the conditions for realizability by considering realizable signings of double occurrence strings. Let $\Sigma$ be an (unsigned) alphabet. Recall that a signed string $v$ over $\Sigma$ is a signing of a string $u \in \Sigma^*$, if $\|v\| = u$. Moreover, if $v$ is realizable, then it is called a **realizable signing** of $u$.

*Example 8.8.* Let $\Sigma = [1,5]$. Then the strings $v_1 = 2\,3\,\overline{2}\,4\,\overline{3}\,5\,4\,5$ and $v_2 = \overline{2}\,3\,2\,\overline{4}\,3\,5\,4\,5$ are both signings of the double occurrence string $u = 2\,3\,2\,4\,3\,5\,4\,5$. Notice that $v_1$ is realistic ($v_1 = \varrho_5(M_2\overline{M}_1\overline{M}_3M_5M_4)$), but $u$ is not. Also, it is easy to see that the signing $v_2$ is not realistic. □

By definition every signing of a double occurrence string yields a legal string. We shall now study the problem of determining which double occurrence strings have realistic signings.

For a double occurrence string $w = a_1a_2\ldots a_{2n}$, we define a vertex- and edge-labelled graph $A_w = (V, E, \varepsilon, \ell, h)$ as follows:

- The set of vertices is $V = \{1, 2, \ldots, 2n\}$ with $\ell(i) = a_i$ for each $i \in [1, 2n]$.
- The set of edges is $E = E_0 \cup E_1$, where

$$E_0 = \{\, e_{ij} \mid a_i = a_j,\ 1 \le i < j \le 2n \,\},$$
$$E_1 = \{\, e'_{ij} \mid j - i = 1,\ 1 \le i < 2n \,\},$$

with $\varepsilon(e_{ij}) = \{i, j\}$ if $e_{ij} \in E_0$, and $\varepsilon(e'_{ij}) = \{i, j\}$ if $e'_{ij} \in E_1$. The labels of the edges are specified by

$$h(e) = \begin{cases} 0, & \text{if } e \in E_0, \\ 1, & \text{if } e \in E_1. \end{cases}$$

If $j - i = 1$ and $a_i = a_j$ both hold, then there are two edges ($e_{ij}$ and $e'_{ij}$) between $i$ and $j$, one edge of each label 0 and 1.

Recall that a path $e_1 e_2 \ldots e_k$ in the graph $A_w$ is alternating if $h(e_{2i+1}) = 0$ and $h(e_{2(i+1)}) = 1$ for each $i$, and it is alternating Hamiltonian if it visits every vertex of the graph exactly once. Recall also that $y \overset{i}{\to} z$ describes the orientation $(y, z)$ of an edge $e$ with $\varepsilon(e) = \{x, y\}$ and $h(e) = i$. In a drawing of the graph $A_w$ (Fig. 8.1) a solid line represents the color 1 and a dashed line the color 0. We sometimes write the labels of the vertices beneath them.

*Example 8.9.* Let $w = 2\,3\,5\,4\,6\,4\,6\,3\,2\,5$ be a double occurrence string over $\Delta_6$. Then the graph $A_w$ is drawn in Fig. 8.1. The path

$$1 \overset{0}{\to} 9 \overset{1}{\to} 10 \overset{0}{\to} 3 \overset{1}{\to} 4 \overset{0}{\to} 6 \overset{1}{\to} 5 \overset{0}{\to} 7 \overset{1}{\to} 8 \overset{0}{\to} 2$$

is an alternating Hamiltonian path of $A_w$.                                □

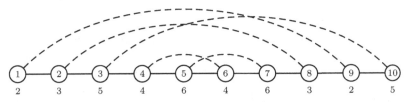

**Fig. 8.1.** The graph $A_w$ in Example 8.9. A *solid line* represents the color 1 and a *dashed line* the color 0. The labels of the vertices are written beneath them.

**Theorem 8.2.** *A double occurrence string $w$ has a realizable signing if and only if the graph $A_w$ has an alternating Hamiltonian path.*

*Proof.* In the following, for simplicity, we shall omit the index $\kappa$ from the function $\varrho_\kappa$.

(1) We show first by induction on the length of the double occurrence strings that every string $w$ that has a realizable signing $w'$ (and thus there exists an isomorphism $\tau$ such that $\tau(w')$ is realistic) has an alternating Hamiltonian path that starts from the vertex $\varrho(M_1) = 2$ corresponding to the beginning marker, and ends in the vertex $\varrho(M_\kappa) = \kappa$ corresponding to the end marker of the realistic legal string $\tau(w')$.

Let $w$ be a double occurrence string of length $2k$ over the alphabet $\Sigma$, and assume that $w$ has a realizable signing $w' \in \Sigma^{\maltese}$. Hence there exists an isomorphism $\tau \colon \Sigma^{\maltese} \to \Delta^{\maltese}_{k+1}$ such that $\tau(w')$ is realistic. Clearly, the original string $w$ and its image $\tau(w)$ have the same graph, $A_w = A_{\tau(w)}$, and therefore we can assume without loss of generality that $w = \tau(w)$. In particular, $w$ is over $\Delta_{k+1}$, and $w'$ is a realistic signing of $w$.

For each $i \in [2, k+1]$, let $x_{i1}, x_{i2} \in [1, 2k]$ be the vertices of $A_w$ such that $\ell(x_{i1}) = i = \ell(x_{i2})$ and $x_{i1} < x_{i2}$.

Now $w = w_1(k+1)w_2(k+1)w_3$ for some $w_1, w_2, w_3 \in \Delta_k^*$. The string $v = w_1 w_2 w_3$ has a realistic signing $v'$ obtained from the signing $w'$ by removing the two occurrences of the elements from $\{k+1, \overline{k+1}\}$. By the induction hypothesis, $v'$ has an alternating Hamiltonian path from $x_{2t}$ to $x_{kr}$, where $t, r \in \{1, 2\}$, and $x_{2t}$ is the position of the beginning marker $\varrho(M_1)$ and $x_{kr}$ of the end marker $\varrho(M_k)$ (of $v'$). Since $w'$ is realistic, the occurrence of this $k = \varrho(M_k)$ is part of the substring $k(k+1)$ or $(\overline{k+1})\,\overline{k}$ in $w'$. It is now obvious that we have an alternating Hamiltonian path

$$ x_{2t} \xrightarrow{0} \ldots \xrightarrow{0} x_{kr} \xrightarrow{1} x_{(k+1)r'} \xrightarrow{0} x_{(k+1)r''} $$

in $A_w$, where $\{r', r''\} = \{1, 2\}$, and $x_{(k+1)r''}$ is the position of the end marker $k+1 = \varrho(M_{k+1})$ in $w'$. This proves the claim in the present direction.

(2) In the other direction, suppose that $w$ is a double occurrence string such that the graph $A_w$ does have an alternating Hamiltonian path. Let the Hamiltonian path be

$$ x_{11} \xrightarrow{0} x_{12} \xrightarrow{1} \ldots \xrightarrow{1} x_{i1} \xrightarrow{0} x_{i2} \xrightarrow{1} \cdots \rightarrow x_{k1} \xrightarrow{0} x_{k2}, $$

where again $\ell(x_{i1}) = \ell(x_{i2})$, for each $i = 1, 2, \ldots, k$, so that the edge $\{x_{i1}, x_{i2}\}$ has color 0. Here $\{x_{i1}, x_{i2} \mid i = 1, 2, \ldots, k\} = [1, 2k]$. Define the morphism $\tau$ by $\tau(\ell(x_{i1})) = i + 1$, and define the signing as follows: if $x_{i2} > x_{(i+1)1}$ (and so $x_{i2} = x_{(i+1)1} + 1$) in the edge $\{x_{i2}, x_{(i+1)1}\}$ of color 1, then sign both $\tau(\ell(x_{i2}))$ and $\tau(\ell(x_{(i+1)1}))$; otherwise $x_{i2} = x_{(i+1)1} - 1$, and in this case, $\tau(\ell(x_{i2}))$ and $\tau(\ell(x_{(i+1)1}))$ are left unsigned. By the construction, the so-obtained string is realistic, and the claim follows from this. □

*Example 8.10.* We illustrate the previous theorem and its proof by continuing Example 8.9. In Fig. 8.2 we are given the alternating Hamiltonian path

$$ 1 \xrightarrow{0} 9 \xrightarrow{1} 10 \xrightarrow{0} 3 \xrightarrow{1} 4 \xrightarrow{0} 6 \xrightarrow{1} 5 \xrightarrow{0} 7 \xrightarrow{1} 8 \xrightarrow{0} 2 $$

of the graph $A_w$. The labels beneath the vertices are now obtained by following the Hamiltonian path as in the previous proof. The signing of the proof produces the realistic legal string $2\,6\,3\,4\,\overline{5}\,\overline{4}\,5\,6\,2\,3$. □

The proof of Theorem 8.2 yields the following procedure to determine whether a given legal string is realizable. Let $v$ be a legal string of length $2m$ over an alphabet $\Sigma$. We consider the graph $A_v = A_{v'}$, where $v' = \|v\|$ is obtained from $v$ by removing the bars. Let $x$ be a vertex of $A_v$. Define an **induced alternating path** $\text{alt}_v(x)$ *of $v$ from $x$* as follows:

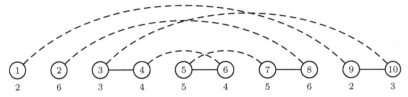

**Fig. 8.2.** Alternating Hamiltonian path of Example 8.10. The path starts from vertex 1, and the edges alternate in their colors. Each vertex is along the path.

(1) Start with the edge $x_1 \xrightarrow{0} x_2$, where $x_1 = x$ and $\ell(x_2) = \ell(x_1)$. Then $\mathrm{alt}_v(x)$ is the maximum path obtained by iterating the following step (2).

(2) Suppose the alternating path $x_1 \xrightarrow{0} x_2 \xrightarrow{1} \ldots \xrightarrow{1} x_{i-1} \xrightarrow{0} x_i$ has been constructed so that the color of the edge $\{x_{i-1}, x_i\}$ is 0.

    (2.1) If the label of $x_i$ is signed in $v$, then the next edges are $\{x_i, x_i - 1\}$ (of color 1) and $\{x_i - 1, x_{i+1}\}$ (of color 0), where $\ell(x_{i+1}) = \ell(x_i - 1)$.

    (2.2) If the label of $x_i$ is not signed in $v$, then the next edges are $\{x_i, x_i + 1\}$ (of color 1) and $\{x_i + 1, x_{i+1}\}$ (of color 0), where $\ell(x_{i+1}) = \ell(x_i + 1)$.

Notice that after the vertex $x$ is chosen, the induced alternating path $\mathrm{alt}_v(x)$ is well defined, that is, in every step the next edge is uniquely determined. The following result is a corollary to Theorem 8.2.

**Theorem 8.3.** *Let $v$ be a legal string. Then $v$ is realizable if and only if there exists a vertex $x$ of $A_v$ such that the induced alternating path $\mathrm{alt}_v(x)$ is an alternating Hamiltonian path of $A_v$.*

## Notes on References

- Double occurrence strings are well known in the literature on circle graphs, see Bouchet [5] and de Fraysseix [13], as well as, in the literature on self-crossing plane curves, see de Fraysseix and Ossona de Mendez [14].
- Realistic legal strings in relation to gene assembly of ciliates were first studied by Ehrenfeucht et al. [20].
- Theorem 8.1 was proved by Ehrenfeucht et al. [16, 17].
- The characterization problem of realizable signings of strings was first considered in Harju and Rozenberg in [26], where the graph $A_w$ was introduced in this context. It is related to the breakpoint graphs of signed permutations, see Hannenhalli and Pevzner [24]. Theorem 8.2, as well as Theorem 8.3, was proved in [26]. We return to this topic in Chap. 10.

# 9

# String Pointer Reduction System

In Chap. 8 we formalized the MDS structure of micronuclear and intermediate genes through legal strings. In this chapter we shall formalize the gene assembly process in the framework of legal strings. Thus, we shall now formalize the three molecular operations *ld*, *hi*, and *dlad* as operations on (rewriting rules for) legal strings, obtaining string pointer reduction systems. Although the transition from MDS descriptors to legal strings represents an increase in abstraction (we only preserve in legal strings the sequence of pointers), we prove that string pointer reduction systems are equivalent to MDS descriptor pointer reduction systems as far as the strategies for gene assembly are concerned.

## 9.1 Assembly Operations on Strings

Recall that each realistic MDS descriptor $\delta = (p_1, q_1) \ldots (p_m, q_m)$ is obtained from a realistic arrangement $\alpha = N_1 N_2 \ldots N_\kappa$ as the image $\psi_\kappa(\alpha) = \delta$, where $\psi_\kappa$ is defined in Chap. 6; see p. 70. For each realistic MDS descriptor $\delta \in \Gamma_\kappa^\maltese$, we associate a realistic legal string $w_\delta$, which is obtained from the MDS descriptor $\delta$ by simply writing the sequence of the pointers and deleting the markers ($e$ or $\overline{e}$, and $b$ or $\overline{b}$) and the parentheses, see Chap. 8. Therefore, in terms of the function $\mathrm{rem}_\kappa$ (see p. 84), we have

$$w_\delta = \mathrm{rem}_\kappa(\delta).$$

*Example 9.1.* The legal string of the following realistic MDS descriptor $\delta = (5, e)(b, 2)(2, 3)(\overline{5}, \overline{4})(3, 4)$ is equal to

$$w_\delta = 5\,2\,2\,3\,\overline{5}\,\overline{4}\,3\,4 = \varrho_5(M_5 M_1 M_2 \overline{M}_4 M_3),$$

where $M_5 M_1 M_2 \overline{M}_4 M_3$ is a micronuclear arrangement. Recall from Chap. 5 that the *p*-interval of a legal string is the substring "sandwiched" by the two occurrences from the pointer set $\mathbf{p} = \{p, \bar{p}\}$, including the two occurrences of $p$. Hence the 2-interval of $w_\delta$ is the substring $22$, the 3-interval of $w_\delta$ is $3\,\bar{5}\,\bar{4}\,3$, the 4-interval is $\bar{4}\,3\,4$, and the 5-interval is $5\,2\,2\,3\,\bar{5}$. Thus, e.g., pointer 3 overlaps in $w_\delta$ with both 4 and 5, but not with 2.    □

We shall now introduce the **string pointer reduction system** as a formal system modelling the transformation of pointers during the gene assembly process. This system consists of three sets of reduction rules operating on legal strings, corresponding to the micronuclear operations ld, hi, and dlad.

In the following we shall consider only strings of pointers, that is, the strings are from $\Delta_\kappa^{\bigstar}$ for some $\kappa \geq 2$. Each of the rules below is a function (or an operation) that maps legal strings to legal strings.

- The **string negative rule** $\mathsf{snr}_p$ for a pointer $p \in \Pi_\kappa$ ($\kappa \geq 2$) is applicable to a legal string of the form $u = u_1 p p u_2$, where $u_1, u_2 \in \Delta_\kappa^{\bigstar}$. The result of this operation is

$$\mathsf{snr}_p(u_1 p p u_2) = u_1 u_2. \tag{9.1}$$

  Let

$$\mathsf{Snr} = \{\mathsf{snr}_p \mid p \in \Pi_\kappa,\ \kappa \geq 2\}$$

  be the set of all string negative rules.

- The **string positive rule** $\mathsf{spr}_p$ for a pointer $p \in \Pi_\kappa$ ($\kappa \geq 2$) is applicable to a legal string of the form $u = u_1 p u_2 \bar{p} u_3$, where $u_1, u_2, u_3 \in \Delta_\kappa^{\bigstar}$. The result of this operation is

$$\mathsf{spr}_p(u_1 p u_2 \bar{p} u_3) = u_1 \bar{u}_2 u_3. \tag{9.2}$$

  Let

$$\mathsf{Spr} = \{\mathsf{spr}_p \mid p \in \Pi_\kappa,\ \kappa \geq 2\}$$

  be the set of all string positive rules.

- The **string double rule** $\mathsf{sdr}_{p,q}$ for pointers $p, q \in \Pi_\kappa$ ($\kappa \geq 2$) with $\mathbf{p} \neq \mathbf{q}$ is applicable to a legal string of the form $u = u_1 p u_2 q u_3 p u_4 q u_5$, where $u_i \in \Delta_\kappa^{\bigstar}$ for each $i$. The result of this operation is

$$\mathsf{sdr}_{p,q}(u_1 p u_2 q u_3 p u_4 q u_5) = u_1 u_4 u_3 u_2 u_5. \tag{9.3}$$

  Let

$$\mathsf{Sdr} = \{\mathsf{sdr}_{p,q} \mid p, q \in \Pi_\kappa,\ \kappa \geq 2\}$$

  be the set of all string double rules.

We shall also use the following more graphical notations for the applications of the operations $\mathsf{snr}_p$, $\mathsf{spr}_p$, and $\mathsf{sdr}_{p,q}$ to a legal string $u$ (assuming that the operations are indeed applicable to $u$):

$$u \xrightarrow{\mathsf{snr}_p} \mathsf{snr}_p(u), \quad u \xrightarrow{\mathsf{spr}_p} \mathsf{spr}_p(u), \quad u \xrightarrow{\mathsf{sdr}_{p,q}} \mathsf{sdr}_{p,q}(u).$$

A composition $\varphi = \varphi_n \ldots \varphi_1$ of operations from $\mathsf{Snr} \cup \mathsf{Spr} \cup \mathsf{Sdr}$ is a **string reduction** of $u$, if $\varphi$ is applicable to $u$. Also, $\varphi$ is **successful** for $u$ if $\varphi(u) = \Lambda$, the empty string.

Note that we have only three types of rules for the legal strings – one rule for each case corresponding to *ld* (snr), *hi* (spr), and *dlad* (sdr). This is to be contrasted to the MDS descriptors for which we had altogether 11 cases of rules. Therefore the string pointer reduction system is considerably simpler than the MDS pointer reduction system of Chap. 7.

*Example 9.2.* (1) The string $u = 5\,2\,2\,3\,\overline{5}\,\overline{4}\,3\,4$ is legal. The rule $\mathsf{snr}_2$ is applicable to $u$, and $\mathsf{snr}_2(u) = 5\,3\,\overline{5}\,\overline{4}\,3\,4$. Also, the rules $\mathsf{spr}_5$ and $\mathsf{spr}_{\overline{4}}$ are applicable to $u$. For these rules, we have $\mathsf{spr}_5(w) = \overline{3}\,\overline{2}\,\overline{2}\,4\,3\,4$ and $\mathsf{spr}_{\overline{4}}(w) = 5\,2\,2\,3\,\overline{5}\,\overline{3}$. Notice that the rule $\mathsf{spr}_4$ is *not applicable* to $u$, since the string $4\,\overline{4}$ is not a scattered substring of $u$. For $\varphi = \mathsf{snr}_5\,\mathsf{spr}_3\,\mathsf{spr}_{\overline{4}}\,\mathsf{snr}_2$, we have $\varphi(u) = \mathsf{snr}_5\,\mathsf{spr}_3\,\mathsf{spr}_{\overline{4}}(5\,3\,\overline{5}\,\overline{4}\,3\,4) = \mathsf{snr}_5\,\mathsf{spr}_3(5\,3\,\overline{5}\,\overline{3}) = \mathsf{snr}_5(5\,5) = \Lambda$, or

$$5\,2\,2\,3\,\overline{5}\,\overline{4}\,3\,4 \xrightarrow{\ \mathsf{snr}_2\ } 5\,3\,\overline{5}\,\overline{4}\,3\,4 \xrightarrow{\ \mathsf{spr}_{\overline{4}}\ } 5\,3\,\overline{5}\,\overline{3}$$
$$\xrightarrow{\ \mathsf{spr}_3\ } 5\,5 \xrightarrow{\ \mathsf{snr}_5\ } \Lambda,$$

and hence $\varphi$ is successful for $u$.

(2) The rules $\mathsf{spr}_2$ and $\mathsf{sdr}_{4,5}$ are applicable to the legal string $w = 2\,4\,\overline{2}\,\overline{3}\,\overline{6}\,5\,6\,4\,3\,5$. We have $\mathsf{spr}_2(w) = \overline{4}\,3\,\overline{6}\,5\,6\,4\,3\,5$ and $\mathsf{sdr}_{4,5}(w) = 2\,3\,6\,\overline{2}\,3\,\overline{6}$. The composition $\mathsf{snr}_2\,\mathsf{spr}_3\,\mathsf{spr}_{\overline{6}}\,\mathsf{sdr}_{4,5}$ is successful for $w$:

$$2\,4\,\overline{2}\,\overline{3}\,\overline{6}\,5\,6\,4\,3\,5 \xrightarrow{\ \mathsf{sdr}_{4,5}\ } 2\,3\,6\,\overline{2}\,3\,\overline{6} \xrightarrow{\ \mathsf{spr}_{\overline{6}}\ } 2\,3\,\overline{3}\,2 \xrightarrow{\ \mathsf{spr}_3\ } 2\,2 \xrightarrow{\ \mathsf{snr}_2\ } \Lambda. \qquad \square$$

## 9.2 Equivalence to Descriptor Pointer Reduction System

We show in this section that there is a one-to-one correspondence between the operations in Ld, Hi, and Dlad on realistic MDS descriptors and the operations in Snr, Spr and Sdr on legal strings.

### 9.2.1 Ld and Snr

We consider first the string negative rule and its correspondence with the operations in Ld for MDS descriptors. Recall from p. 79 that we allow only

$$\mathsf{ld}_p((p,m)(m',p)) = (m',m), \quad (m,m' \in \mathcal{M}) \qquad (\ell 2')$$

as the boundary rules for the MDS descriptors.

**Lemma 9.1.** *Let $\delta \in \Gamma_\kappa^{\maltese}$ be a realistic MDS descriptor, and let $p \in \Pi_\kappa$ be a pointer. If $\mathsf{ld}_p$ is applicable to $\delta$, then $\mathsf{snr}_p$ is applicable to $w_\delta$. In this case, $w_{\mathsf{ld}_p(\delta)} = \mathsf{snr}_p(w_\delta)$.*

*Proof.* If $\mathsf{Id}_p$ is applicable to $\delta$, then this application is either simple or boundary. In both cases $pp$ is a substring of $w_\delta$, and so $\mathsf{snr}_p$ is applicable to $w_\delta$. Assume that $\delta$ is in $\Gamma_\kappa^{\maltese}$, and let $\delta' = \mathsf{Id}_p(\delta)$.

Let first $\delta = \delta_1(q,p)(p,r)\delta_2$. Then $\delta' = \delta_1(q,r)\delta_2$, and hence in the legal string $w_{\delta'} = \mathrm{rem}_\kappa(\delta') = \mathrm{rem}_\kappa(\delta_1)\,\mathrm{rem}_\kappa(q,r)\,\mathrm{rem}_\kappa(\delta_2)$ the sequence of the remaining pointers is the same as in $w_\delta = \mathrm{rem}_\kappa(\delta)$, that is, $w_{\delta'}$ equals $w_\delta$ with $pp$ removed. On the other hand, applying $\mathsf{snr}_p$ to $w_\delta$ removes both occurrences of $p$ in $w_\delta$ yielding $w_{\delta'}$, as required.

Let then $\delta = (p,m)(m',p)$, where $m, m' \in \mathcal{M}$. Then $\delta' = (m',m)$, and so $w_{\delta'} = \Lambda$. Also, $\mathsf{snr}_p(w_\delta) = \mathsf{snr}_p(pp) = \Lambda$, and this proves the lemma.      □

**Lemma 9.2.** *Let $\delta \in \Gamma_\kappa^{\maltese}$ be a realistic MDS descriptor, and let $p \in \Pi_\kappa$ be a pointer. If $\mathsf{snr}_p$ is applicable to $w_\delta$, then $\mathsf{Id}_p$ is applicable to $\delta$. In this case, $\mathsf{snr}_p(w_\delta) = w_{\mathsf{Id}_p(\delta)}$.*

*Proof.* Suppose that $\mathsf{snr}_p$ is applicable to $w_\delta$. Then $pp$ is a substring of $w_\delta$, and hence only markers may be present between the two occurrences of $p$ in the MDS descriptor $\delta$. Therefore $\delta = \delta_1 x y \delta_2$, where $x, y \in \Gamma_\kappa \cup \overline{\Gamma}_\kappa$ are the pairs containing $p$ and $\delta_1, \delta_2 \in \Gamma_\kappa^{\maltese}$. (It follows from Corollary 6.2 and the definition of MDS descriptors that there must be exactly *two* such pairs $x, y$.) We have three cases to be considered:

*Case 1.* There are no markers in $\delta$ between the two occurrences of $p$. Since $\delta$ is a realistic MDS descriptor, necessarily $\delta = \delta_1(q_1,p)(p,q_2)\delta_2$ for some $q_1, q_2 \in \Pi_\kappa \cup \mathcal{M}$. Clearly, $\mathsf{Id}_p$ is applicable to $\delta$.

*Case 2.* There is one marker in $\delta$ between the two occurrences of $p$. Now either $\delta = \delta_1(p,m)(p,r)\delta_2$ or $\delta = \delta_1(r,p)(m,p)\delta_2$ for some $r \in \Pi_\kappa \cup \mathcal{M}$ and $m \in \mathcal{M}$. However, in these cases either both occurrences of $p$ are left, or both occurrences of $p$ are right, which is not possible by Corollary 6.1.

*Case 3.* There are two markers $m, m' \in \mathcal{M}$ in $\delta$ between the two occurrences of $p$. We have three subcases here.

*Case 3.1.* $\delta = (p,m)(m',p)$ – so that $\delta_1\delta_2 = \Lambda$. Clearly, $\mathsf{Id}_p$ is applicable to $\delta$, yielding $(m',m)$.

*Case 3.2.* $\delta = \delta_1(p,m)(m',p)\delta_2$ with $\delta_1\delta_2 \neq \Lambda$. In this case, $(p,m)(m',p)$ is a realistic MDS descriptor that is a proper scattered substring of $\delta$. Since $\delta$ is a realistic MDS descriptor, Corollary 6.1 implies that this case is not possible.

*Case 3.3.* $\delta = \delta_1(q_1,p)(m,m')(p,q_2)\delta_2$ for some $q_1, q_2 \in \Pi_\kappa$. This contradicts Corollary 6.2, and so this case is also not possible.

Hence, only Case 1 and Case 3.1 are possible, and in both cases $\mathsf{Id}_p$ is applicable. The second claim follows from Lemma 9.1.      □

From Lemma 9.1 and Lemma 9.2, we obtain intertranslatability of the operations $\mathsf{Id}_p$ on MDS descriptors and the string negative reduction rules $\mathsf{snr}_p$ on legal strings.

**Corollary 9.1.** *Let $\delta \in \Gamma_\kappa^{\maltese}$ be a realistic MDS descriptor and $p \in \Pi_\kappa$ be a pointer. The rule $\mathsf{ld}_p$ is applicable to $\delta$ if and only if $\mathsf{snr}_p$ is applicable to $w_\delta$. In this case,*

$$\mathsf{snr}_p(w_\delta) = w_{\mathsf{ld}_p(\delta)}.$$

### 9.2.2  Hi and Spr

We now consider the correspondence between the string positive rule and the hi-rule of MDS descriptors.

**Lemma 9.3.** *Let $\delta \in \Gamma_\kappa^{\maltese}$ be a realistic MDS descriptor and let $p \in \Pi_\kappa$ be a pointer. If $\mathsf{hi}_p$ is applicable to $\delta$, then $\mathsf{spr}_p$ is applicable to $w_\delta$. In this case, $\mathsf{spr}_p(w_\delta) = w_{\mathsf{hi}_p(\delta)}$.*

*Proof.* If $\mathsf{hi}_p$ is applicable to $\delta$, then $p$ and $\overline{p}$ occur in $\delta$ in this order. By Corollary 6.1, they are either both left or both right in $\delta$. Since $\mathsf{hi}_p$ is applicable to $\delta$, $p\overline{p}$ is a scattered substring of $w_\delta$, and so the string reduction rule $\mathsf{spr}_p$ is applicable to $w_\delta$.

Now either (9.4) or (9.5) of the following cases holds:

$$\delta = \delta_1(p, q_1)(q_2, q_3) \ldots (q_{2i}, q_{2i+1})(\overline{p}, r)\delta_2, \tag{9.4}$$

$$\delta = \delta_1(r, p)(q_1, q_2) \ldots (q_{2i-1}, q_{2i})(q_{2i+1}, \overline{p})\delta_2, \tag{9.5}$$

where $q_1, \ldots, q_{2i+1}, r \in \Pi_\kappa \cup \mathcal{M}$ and $\delta_1, \delta_2 \in \Gamma_\kappa^{\maltese}$.

In Case (9.4), $\mathsf{hi}_p(\delta) = \delta_1(\overline{q}_{2i+1}\overline{q}_{2i}) \ldots (\overline{q}_3, \overline{q}_2)(\overline{q}_1, r)\delta_2$, and in Case (9.5), $\mathsf{hi}_p(\delta) = \delta_1(r, \overline{q}_{2i+1})(\overline{q}_{2i}, \overline{q}_{2i-1}) \ldots (\overline{q}_2, \overline{q}_1)\delta_2$. It then follows that $\mathsf{spr}_p(w_\delta) = w_{\mathsf{hi}_p(\delta)}$. □

**Lemma 9.4.** *Let $\delta \in \Gamma_\kappa^{\maltese}$ be a realistic MDS descriptor and let $p \in \Pi_\kappa$ be a pointer. If $\mathsf{spr}_p$ is applicable to $w_\delta$, then $\mathsf{hi}_p$ is applicable to $\delta$. In this case, $\mathsf{spr}_p(w_\delta) = w_{\mathsf{hi}_p(\delta)}$.*

*Proof.* Since $\mathsf{spr}_p$ is applicable to $w_\delta$, $p\overline{p}$ is a scattered substring of $w_\delta$. Moreover, by Corollary 6.1, either both $p$ and $\overline{p}$ are left pointers in $\delta$, or they are both right pointers in $\delta$. In both cases, the rule $\mathsf{hi}_p$ is applicable to $\delta$.

Now, by Lemma 9.3 we have $\mathsf{spr}_p(w_\delta) = w_{\mathsf{hi}_p(\delta)}$. □

Lemma 9.3 together with Lemma 9.4 yields intertranslatability of the operations $\mathsf{hi}_p$ on realistic MDS descriptors and the string reduction rules $\mathsf{spr}_p$ on legal strings.

**Corollary 9.2.** *Let $\delta \in \Gamma_\kappa^{\maltese}$ be a realistic MDS descriptor and $p \in \Pi_\kappa$ be a pointer. The rule $\mathsf{hi}_p$ is applicable to $\delta$ if and only if $\mathsf{spr}_p$ is applicable to $w_\delta$. In this case,*

$$\mathsf{spr}_p(w_\delta) = w_{\mathsf{hi}_p(\delta)}.$$

### 9.2.3 Dlad and Sdr

Finally we consider the correspondence between the string double rule sdr and the operation dlad on realistic MDS descriptors.

**Lemma 9.5.** *Let $\delta \in \Gamma_\kappa^{\maltese}$ be a realistic MDS descriptor and let $p, q \in \Pi_\kappa$ be pointers such that $\mathbf{p} \neq \mathbf{q}$. If $\mathsf{dlad}_{p,q}$ is applicable to $\delta$, then $\mathsf{sdr}_{p,q}$ is applicable to $w_\delta$. In this case, $\mathsf{sdr}_{p,q}(w_\delta) = w_{\mathsf{dlad}_{p,q}(\delta)}$.*

*Proof.* If $\mathsf{dlad}_{p,q}$ is applicable to $\delta$, then pointer $p$ overlaps with pointer $q$ in $\delta$, and hence $\mathsf{sdr}_{p,q}$ is applicable to $w_\delta$.

Applying $\mathsf{dlad}_{p,q}$ to $\delta$ removes the occurrences of $p$ and $q$, and the sequence between the first occurrences of $p$ and $q$ in $\delta$ is interchanged with the sequence between the second occurrences of $p$ and $q$ in $\delta$. Consequently, $\mathsf{sdr}_{p,q}(w_\delta) = w_{\mathsf{dlad}_{p,q}(\delta)}$. □

Conversely, we have

**Lemma 9.6.** *Let $\delta \in \Gamma_\kappa^{\maltese}$ be a realistic MDS descriptor and let $p, q \in \Pi_\kappa$ be pointers such that $\mathbf{p} \neq \mathbf{q}$. If $\mathsf{sdr}_{p,q}$ is applicable to $w_\delta$, then $\mathsf{dlad}_{p,q}$ is applicable to $\delta$. In this case, $\mathsf{sdr}_{p,q}(w_\delta) = w_{\mathsf{dlad}_{p,q}(\delta)}$.*

*Proof.* If $\mathsf{sdr}_{p,q}$ is applicable to $w_\delta$, then both $p$ and $q$ have two occurrences in $\delta$, a left one and a right one. We consider the following cases according to the occurrences of $p$ and $q$ in $\delta$:

1. The first occurrences of both $p$ and $q$ are left.
2. The first occurrences of both $p$ and $q$ are right.
3. The first occurrence of $p$ is left, and that of $q$ is right.
4. The first occurrence of $p$ is right, and that of $q$ is left.

We prove the claim only for the first case, as the reasoning for the other cases is similar. Assume then that the first occurrence of $p$ in $\delta$ is left, and the first occurrence of $q$ is also left in $\delta$, and then let $w_\delta = w_1 p w_2 q w_3 p w_4 q w_5$ for some strings $w_1, w_2, w_3, w_4, w_5$. Clearly, between the first occurrences of $p$ and $q$ in $\delta$ there must be at least one pointer or marker: $\delta = \delta_1(p, r_1)\delta_2(q, r_2)\delta'$, where $r_1, r_2 \in \Pi_\kappa \cup \mathcal{M}$ and $\delta_1, \delta_2, \delta' \in \Gamma_\kappa^{\maltese}$. Now, if the middle string $w_3$ is nonempty, then there must be at least one pointer or marker between the similar occurrences of the pointers $p$ and $q$ in $\delta$: $\delta = \delta_1(p, r_1)\delta_2(q, r_2)\delta_3(r_3, p)\delta_4(r_4, q)\delta_5$ for some $r_3, r_4 \in \Pi_\kappa \cup \mathcal{M}$ and $\delta_3, \delta_4, \delta_5 \in \Gamma_\kappa^{\maltese}$. On the other hand, if $w_3 = \Lambda$, then either between the first occurrence of $q$ and the second occurrence of $p$ there is a marker, which is a particular case of the previous one, or there is no pointer or marker between them: $\delta = \delta_1(p, r_1)\delta_2(q, p)\delta_4(r_4, q)\delta_5$. In both cases, $\mathsf{dlad}_{p,q}$ is applicable to $\delta$. Indeed, in the former case this is so by the instance (a) in the definition of $\mathsf{dlad}_{p,q}$ (see Sect. 7.1), and in the latter case by the instance (e) in the same definition.

The second part of the claim follows directly from Lemma 9.5. □

Lemma 9.5 together with Lemma 9.6 give intertranslatability of the operations $\mathsf{dlad}_{p,q}$ on realistic MDS descriptors and the string reduction rules $\mathsf{sdr}_{p,q}$ on legal strings.

**Corollary 9.3.** *Let* $\delta \in \Gamma_\kappa^{\maltese}$ *be a realistic MDS descriptor, and let* $p, q \in \Pi_\kappa$ *be pointers such that* $\mathbf{p} \neq \mathbf{q}$. *The rule* $\mathsf{dlad}_{p,q}$ *is applicable to* $\delta$ *if and only if* $\mathsf{sdr}_{p,q}$ *is applicable to* $w_\delta$. *In this case,*

$$\mathsf{sdr}_{p,q}(w_\delta) = w_{\mathsf{dlad}_{p,q}(\delta)}.$$

By combining Corollaries 9.1, 9.2, and 9.3 together with Theorem 7.1, we obtain the universality of the string pointer reduction system:

**Theorem 9.1.** *Each legal string has a successful string reduction.*

We also have the following corollary to Lemma 7.1 using the above results.

**Corollary 9.4.** *Let* $\varphi$ *be a string reduction for a realistic legal string* $v$. *Then the image* $\varphi(v)$ *is also realistic.*

*Proof.* It is sufficient to show the claim for $\varphi \in \mathsf{Snr} \cup \mathsf{Spr} \cup \mathsf{Sdr}$, since the general claim then follows by induction on the number of operations in the composition. Let $v = w_\delta$ for a realistic MDS descriptor $\delta$. By Corollary 9.1, if $\varphi = \mathsf{snr}_p$ for a pointer $p$, then $\mathsf{snr}_p$ is applicable to $w_\delta$ and $\mathsf{snr}_p(w_\delta) = w_{\mathsf{ld}_p(\delta)}$. Hence $\mathsf{snr}_p(w_\delta)$ is realistic, since, by Lemma 7.1, $\mathsf{ld}_p(\delta)$ is realistic, and $w_{\mathsf{ld}_p(\delta)} = \mathsf{rem}_\kappa(\mathsf{ld}_p(\delta))$.

The cases for $\mathsf{spr}_p$ and $\mathsf{sdr}_{p,q}$ are similar, but use Corollaries 9.2 and 9.3, respectively. $\qquad\square$

# Notes on References

- The string pointer reduction system was introduced by Ehrenfeucht et al. [15, 20]. In the first of these articles [20] the operations in Snr and Spr were considered and their correspondence with the operations in Ld and Hi of the MDS descriptors were established. Then in [15] the remaining string operations in Sdr were studied and their relation to operations in Dlad was established.
- The signed permutations over $[1, \kappa]$ are in a natural one-to-one correspondence with the micronuclear arrangements of size $\kappa$. This correspondence is given by: $i \longleftrightarrow M_i$ and $\bar{i} \longleftrightarrow \overline{M}_i$. For instance, the micronucleus version of the actin gene in *Sterkiella nova* (see Example 6.1) is given by the arrangement $M_3 M_4 M_6 M_5 M_7 M_9 \overline{M}_2 M_1 M_8$, which corresponds to the signed permutation $3\,4\,6\,5\,7\,9\,\bar{2}\,1\,8$.
  By Theorem 9.1, every legal string can be reduced to $\Lambda$ by using the three operations snr, spr, and sdr. Such a reduction corresponds to sorting the signed permutation (cyclically) in increasing order.

Now every signed permutation can be sorted by using general inversions (without reference to pointers), that is, by using the operation $xyz \mapsto x\bar{y}z$, where $y$ is an arbitrary substring. It was shown by Hannenhalli and Pevzner [24] (see also Berman and Hannenhalli [4] and Kaplan et al. [31]) that there is a fast algorithm for finding the smallest number of reversals needed to sort any given signed permutation. Interestingly, the corresponding problem of sorting unsigned permutations by reversals is known to be difficult (NP-hard); see Caprara [7].

The string positive rule spr is a special case of the inversion operation. Indeed, in the operation for signed permutations corresponding to spr we require that the inversions are performed between two consecutive integers (of which exactly one is signed) in the signed permutation. Therefore the corresponding operation of spr *cannot sort* all signed permutations.

- General formal language theory aspects of the string operations related to snr, spr and sdr have been considered, e.g., by Daley [8], Daley and Kari [9], Daley et al. [10], and Dassow et al. [11].

# Overlap Graphs

All formalizations of MDS structure of micronuclear or intermediate genes that we have considered until now (viz., micronuclear arrangements, MDS descriptors, and legal strings) were done in the framework of strings. We will move now to a different mathematical framework by formalizing MDS structure of genes as graphs. This transition from strings to graphs is obtained by considering the overlap graphs of legal strings. Each pointer set occurring in a legal string determines the substring of this string delimited by the two occurrences of this pointer set, and the overlap graph of a legal string represents the overlapping structures of all such substrings. In this chapter we introduce the basic notions concerning overlap graphs and study their relationship to (realistic) legal strings.

## 10.1 Overlap Graphs of Legal Strings

Recall that in a simple graph $(V, E)$ each edge $e \in E$ is an unordered pair $\{x, y\}$ of different vertices, i.e., we do not have an endpoint mapping $\varepsilon$ for these graphs.

A **signed graph** $\gamma = (V, E, \sigma)$ consists of a simple graph $(V, E)$ together with a labelling $\sigma \colon V \to \{-, +\}$ of the vertices. Vertices labelled by $+$ are **positive**, and those labelled by $-$ are **negative**. We use $x^{\sigma(x)}$ to denote a vertex $x$ together with its label $\sigma(x)$.

We shall use signed graphs to represent the structure of overlaps of pointers in a legal string as follows. Let $v \in \Delta_\kappa^{\maltese}$ be a legal string of pointers, and let $\mathbf{P}_v$ be the set of those pointer sets $\mathbf{p} = \{p, \bar{p}\}$ such that $p$ occurs in $v$. Then the **overlap graph** of $v$ is the signed graph $\gamma_v = (\mathbf{P}_v, E, \sigma)$ such that

$$\sigma(\mathbf{p}) = \begin{cases} + , & \text{if } p \in \Delta_\kappa \text{ is positive in } v, \\ - , & \text{if } p \in \Delta_\kappa \text{ is negative in } v, \end{cases}$$

and

$$\{\mathbf{p}, \mathbf{q}\} \in E \iff p \text{ and } q \text{ overlap in } v.$$

*Example 10.1.* Consider the legal string $v = 3\,4\,\bar{5}\,2\,\bar{3}\,\bar{2}\,4\,5 \in \Delta_5^{\maltese}$. Then its overlap graph is given in Fig. 10.1. Indeed, pointers 2, 3, and 5 are positive in $v$ (and hence the vertices **2**, **3**, and **5** have sign $+$ in the overlap graph $\gamma_v$), while pointer 4 is negative in $v$ (and the vertex **4** has sign $-$ in $\gamma_v$). The intervals of the pointers in $v$ are

$$v_{(2)} = 2\,\bar{3}\,\bar{2}, \qquad\qquad v_{(3)} = 3\,4\,\bar{5}\,2\,\bar{3},$$
$$v_{(4)} = 4\,\bar{5}\,2\,\bar{3}\,\bar{2}\,4, \qquad v_{(5)} = \bar{5}\,2\,\bar{3}\,\bar{2}\,4\,5,$$

and therefore $O_v(2) = \{2,3\}$, $O_v(3) = \{2,3,4,5\}$, and $O_v(4) = \{3,4,5\} = O_v(5)$. These overlap sets yield the edges of the overlap graph. □

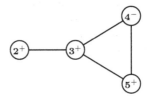

**Fig. 10.1.** The overlap graph of the signed string $v = 3\,4\,\bar{5}\,2\,3\,\bar{2}\,4\,5$. An edge is present between the signed vertices **p** and **q** if and only if $p$ and $q$ overlap in $v$.

*Remark 10.1.* Overlap graphs of double occurrence strings are also known as circle graphs. These graphs have the following geometric interpretation. A **chord diagram** $(\mathcal{C}, V)$ consists of a circle $\mathcal{C}$ in the plane together with a set $V$ of chords. A simple graph $\gamma = (V, E)$ is a **circle graph** if there exists a chord diagram $(\mathcal{C}, V)$ such that $\{x, y\} \in E$ for different $x$ and $y$ if and only if the chords $x$ and $y$ intersect each other. We can see the connection between overlap graphs of legal strings and circle graphs as follows. Consider a circle $\mathcal{C}$ in the plane and write the string $w$ clockwise along $\mathcal{C}$. Then each letter $a$ of $w$ determines a chord the endpoints of which are labelled by $a$. It is clear that two chords intersect if and only if the corresponding letters of $w$ overlap each other. On the other hand, each (signed) circle graph is of the form $\gamma_w$ for a legal string $w$. Indeed, given a chord diagram of a circle graph $\gamma$, label both ends of each chord by the same letter (different letter for each chord), and then read the endpoints of the chords along the circle to obtain a double occurrence string $w$ such that $\gamma = \gamma_w$. □

*Example 10.2.* Consider the legal string $v = 3\,4\,\bar{5}\,2\,\bar{3}\,\bar{2}\,4\,5$ of Example 10.1 the overlap graph of which is given in Fig. 10.1. Then the chord diagram of this example is shown in Fig. 10.2. □

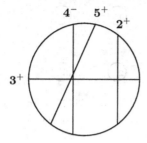

**Fig. 10.2.** The chord diagram of the overlap graph in Example 10.2. A chord represents a vertex of the overlap graph, and two chords intersect if and only if the corresponding vertices are adjacent in the overlap graph.

The mapping $w \mapsto \gamma_w$ of legal strings to overlap graphs is not injective. Indeed, for each legal string $w = w_1 w_2$, we have

$$\gamma_{w_1 w_2} = \gamma_{w_2 w_1}, \quad \text{and} \quad \gamma_w = \gamma_{c(w)},$$

where $c$ is any morphism that selects one element $c(p)$ from $\mathbf{p} = \{p, \bar{p}\}$ for each $p$ (and is faithful for inversion, i.e., $c(\bar{p}) = \overline{c(p)}$). In particular, all conjugates of a legal string $w$ have the same overlap graph. Also, the reversal $w^R$ and the complementation $w^C$ of a legal string $w$ define the same overlap graph as $w$ does.

*Example 10.3.* The following eight legal strings of pointers (in $\Delta_3^*$) have the same overlap graph (given in Fig. 10.3): $2\bar{3}23$, $\bar{3}232$, $232\bar{3}$, $32\bar{3}2$, $\bar{2}3\bar{2}3$, $\bar{3}2\bar{3}2$, $\bar{2}3\bar{2}3$, $3\bar{2}3\bar{2}$.    □

**Fig. 10.3.** The overlap graph of Example 10.3 for eight legal strings on $\Delta_3^*$

*Example 10.4.* All string representations of an overlap graph $\gamma$ need not be obtained from a fixed representation string $w$ by using the operations of conjugation, inversion, and reversing (and by considering isomorphic strings equal). For instance, the strings $v_1 = 23342554$ and $v_2 = 35242453$ define the same overlap graph (Fig. 10.4), while one cannot obtain one of these strings from the other by the above operations. Note, however, that $v_2$ is not realistic. As a matter of fact, we shall prove that such a situation cannot arise if both strings are realistic (see Theorem 10.2).    □

**Fig. 10.4.** The overlap graph of $v_1 = 23342554$ and $v_2 = 35242453$. The signed strings $v_1$ and $v_2$ are not obtainable from each other by the operations of conjugation, reversal, and complementation.

For a simple graph $\gamma$, we denote by $\mathsf{loc}_x(\gamma)$ the graph that is obtained from $\gamma$ by complementing the subgraph induced by the neighborhood $N_\gamma(x)$ of the vertex $x$. Two simple graphs $\gamma$ and $\gamma'$ are **locally equivalent** if there exists a finite sequence $x_1, \ldots, x_k$ of vertices such that $\gamma' = \mathsf{loc}_{x_k} \mathsf{loc}_{x_{k-1}} \ldots \mathsf{loc}_{x_1}(\gamma)$.

*Example 10.5.* Figure 10.5 gives a simple graph $\gamma$ and the locally equivalent graph $\mathsf{loc}_x(\gamma)$.    □

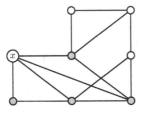 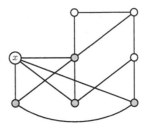

**Fig. 10.5.** The graph $\gamma$ from Example 10.5 and the graph $\mathsf{loc}_x(\gamma)$, which is obtained by complementing the neighborhood of the vertex $x$ (the neighbors of the vertex $x$ are *shaded*).

## 10.2 Realizable Graphs

In Example 8.5 we showed that the legal string $2\,3\,4\,3\,2\,4$ is not realizable. We shall now show that there exist overlap graphs that cannot be "realized" by any micronuclear arrangement.

For the next two proofs we let $A = \{x_1, x_2, x_3\}$ be an alphabet, and we adopt the notation

$$\langle x_i \rangle = x_{i1} x_{i2} x_{i3} x_{i0} x_{i1} x_{i2} x_{i3}.$$

For the proof of the following lemma, recall the definition of the vertex- and edge-labelled graph $A_w$ defined for each double occurrence string $w$ on p. 87.

**Lemma 10.1.** *The conjugates of the double occurrence string*

$$w' = x_{10}x_{20}x_{30}\langle x_1\rangle\langle x_2\rangle\langle x_3\rangle$$

*do not have realizable signings.*

*Proof.* In the following we let the letters $y$ and $y_i$ be variables, and we write

$$\langle y\rangle = y_1y_2y_3y_0y_1y_2y_3. \tag{10.1}$$

Let $u$ be a realizable double occurrence string such that either $u = v_1\langle y\rangle v_2$ or $u = w_2vw_1$, where $\langle y\rangle = w_1w_2$. The specified substring $\langle y\rangle$ or the conjugate $w_2w_1$ of $\langle y\rangle$ in the above will be called the $\langle y\rangle$-**block** of $u$. By Theorem 8.2, there exists an alternating Hamiltonian path of the graph $A_u$. Let $H$ be any such path.

It is straightforward to show by case analysis that if the path $H$ does not start at a position inside the $\langle y\rangle$-block, then $H$ ends at a position inside the $\langle y\rangle$-block, and symmetrically, if $H$ does not end at a position inside the $\langle y\rangle$-block, then $H$ begins at a position inside the $\langle y\rangle$-block. In Fig. 10.6 we illustrate one possibility to travel the substring $\langle y\rangle$ in $H$ for the case $u = v_1\langle y\rangle v_2$. (In the figure, the vertices of $A_u$ are simply represented by their labels.) There the path $H$ visits outside the string $\langle y\rangle$ between the last $y_3$ and the next $z = y_0$. The path ends in $y_0$ in the middle of $\langle y\rangle$.

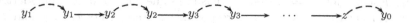

**Fig. 10.6.** A part of the path $H$ in the proof of Lemma 10.1

Therefore the path $H$ must either start or end at a vertex corresponding to an occurrence of a pointer in the $\langle y\rangle$-block. It follows then that any conjugate of a realizable double occurrence string can contain at most two different blocks, a $\langle y\rangle$-block and a $\langle y'\rangle$-block (of the form (10.1)) such that these do not share letters. In particular, no conjugate of the string $w'$ from the statement of the lemma has any realizable signings.    □

As discussed in Sect. 10.1, the mapping $w \mapsto \gamma_w$ from the legal strings to the overlap graphs is not injective. Therefore Lemma 10.1 leaves unanswered the question whether there exists an overlap graph that is not an image $\gamma_w$ of a realizable legal string $w$. We shall demonstrate now that this problem has a positive answer by constructing an (unsigned) overlap graph that is not realizable.

**Theorem 10.1.** *There exists an overlap graph of a double occurrence string that has no signing of the vertices such that the result is an overlap graph of a realistic legal string.*

*Proof.* Let $w' = x_{10}x_{20}x_{30}\langle x_1\rangle\langle x_2\rangle\langle x_3\rangle$ be as stated in Lemma 10.1. The overlap graph $\gamma = \gamma_{w'}$ of $w'$ is given in Fig. 10.7. We will show that this graph is different from $\gamma_w$ for all realizable double occurrence strings $w$.

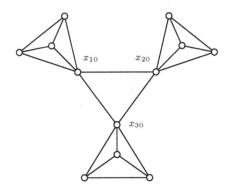

**Fig. 10.7.** The overlap graph of a double occurrence string from the proof of Theorem 10.1. This graph has no realizable signing.

Assume to the contrary that there exists a realizable double occurrence string $w$ such that $\gamma_w = \gamma$. Hence $w$ has the same pairs of overlapping letters as $w'$. For each $i$, the vertices corresponding to the letters $x_{i1}$, $x_{i2}$ and $x_{i3}$ are adjacent to each other in $\gamma$, and they have the same neighborhood in the rest of $\gamma$ (viz., the vertex corresponding to $x_{i0}$). Therefore, without loss of generality, we can assume that they occur in $w$ in the order $x_{i1}, x_{i2}, x_{i3}$, i.e.,

$$w = v_{i0}x_{i1}v_{i1}x_{i2}v_{i2}x_{i3}v_{i3}x_{i1}v_{i4}x_{i2}v_{i5}x_{i3}v_{i6}$$

for some substrings $v_{ij}$ for each $i$.

For each $j \in [0,3]$, the letters $x_{j0}, x_{j1}, x_{j2}, x_{j3}$ overlap with each other in $w$, and they do not overlap with $x_{i1}, x_{i2}, x_{i3}$ for any $i$ with $i \neq j$. Therefore, for each $i \neq j$, either

(a)  all occurrences of $x_{j0}, x_{j1}, x_{j2}, x_{j3}$ are in one string $v_{it}$ (for some $t \in [1,5]$), or

(b)  all occurrences of $x_{j0}, x_{j1}, x_{j2}, x_{j3}$ are in $v_{i0}v_{i6}$.

We show that there are indices $i$ and $j$ such that Case (a) holds. Indeed, there is an index $i$ such that $\langle x_i\rangle$ is not a substring of $w$, since otherwise there would be three disjoint substrings in $w$ of the form (10.1), which is not possible by the proof of Lemma 10.1. Now if the occurrences of $x_{j0}, x_{j1}, x_{j2}, x_{j3}$, for both indices $j$ with $j \neq i$, are all in $v_{i0}v_{i6}$, then $v_{ir} = \Lambda$ or $v_{ir} = x_{i0}$ for each $r \in [1,5]$. Also, one occurrence of $x_{i0}$ must be in $v_{i0}v_{i6}$ (since $x_{i0}$ overlaps with $x_{j0}$), and therefore the second occurrence of $x_{i0}$ must be in $v_{i3}$ (since $x_{i0}$

overlaps with $x_{i1}, x_{i2}$ and $x_{i3}$). However, this means that $w = v_{i0}\langle x_i \rangle v_{i6}$, and so $\langle x_i \rangle$ is a substring of $w$. This is a contradiction.

Let then $i$ and $j$ be indices for which Case (a) holds. Now, exactly one occurrence of $x_{i0}$ lies in $v_{it}$, since $x_{i0}$ overlaps with $x_{j0}$. This occurrence cannot be in an $x_{jr}$-interval of $w$ for any $r \in [1,3]$, since $x_{i0}$ does not overlap with $x_{jr}$ (for $r \in [1,3]$) and the other occurrence of $x_{i0}$ is not in $v_{it}$. Hence this occurrence of $x_{i0}$ is in $v_{j0}v_{j6}$ as well as in $v_{it}$.

Since $x_{j0}$ overlaps with $x_{i0}$ and with each $x_{jr}$ for $r \in [1,3]$, we have

$$\langle x_j \rangle x_{i0} x_{j0} \quad \text{or} \quad x_{j0} x_{i0} \langle x_j \rangle \text{ is scattered substring in } v_{it}. \tag{10.2}$$

Let $k \notin \{i, j\}$ be the remaining index from $\{1, 2, 3\}$. Since $x_{k0}$ overlaps with $x_{j0}$, an occurrence of $x_{k0}$ lies in the $x_{j0}$-interval, and thus it lies in $v_{it}$. Similarly as for $j$, we can then show that (10.2) also holds for $k$:

$$\langle x_k \rangle x_{i0} x_{k0} \quad \text{or} \quad x_{k0} x_{i0} \langle x_k \rangle \text{ is scattered substring in } v_{it}. \tag{10.3}$$

Now $x_{k0}$ does not occur in $v_{jr}$ for any $r \in [1, 5]$, because otherwise (see Cases (a) and (b) in the beginning of the proof) both occurrences of $x_{i0}$ are in $v_{jr}$, but we have shown that one occurrence of $x_{i0}$ is in $v_{j0}v_{j6}$. Therefore $\langle x_j \rangle$ is a substring of $v_{it}$, and, consequently, $\langle x_k \rangle$ is also a substring of $v_{it}$. Hence a conjugate of $w$ has three disjoint substrings of the form $\langle y \rangle$. This is a contradiction by Lemma 10.1, and therefore the proof of the theorem is completed.                                                                                      □

*Example 10.6.* The simplest form of the counterexample given by the previous proof is the string $2\,3\,4\,5\,6\,7\,2\,5\,6\,7\,8\,9\,10\,3\,8\,9\,10\,11\,12\,13\,4\,11\,12\,13$.                           □

## 10.3 The Overlap Equivalence Problem

Examples 10.3 and 10.4 lead naturally to the *overlap equivalence problem* for realistic strings: when do two realistic legal strings $u$ and $v$ have the same overlap graph, i.e., $\gamma_u = \gamma_v$? To solve this problem we begin by giving a characterization of legal substrings of realistic strings.

For a signed string $v \in \Delta_\kappa^\maltese$, let

$$v_{\min} = \min(\text{dom}(v)), \quad \text{and} \quad v_{\max} = \max(\text{dom}(v)).$$

Thus, $\text{dom}(v) \subseteq [v_{\min}, v_{\max}]$.

*Example 10.7.* The string $v = \overline{4}\,8\,3\,4\,\overline{8}\,3$ is legal. We have $\text{dom}(v) = \{3, 4, 8\}$, and hence $v_{\min} = 3$ and $v_{\max} = 8$.                                                      □

**Lemma 10.2.** *Let $u$ be a realistic legal string with $\text{dom}(u) = [2, \kappa]$, and let $v$ be a legal substring of $u$. Then either $\text{dom}(v) = [p, q]$ or $\text{dom}(v) = [2, \kappa] \setminus [p, q]$, for some $p$ and $q$ with $p \le q$. Moreover, $v$ has a realizable conjugate.*

*Proof.* Let $u = u_1vu_2$. We can assume that $v$ is a proper substring of $u$, i.e., $u_1$ and $u_2$ are not both empty strings.

(a) Assume that there exists a pointer $p$ with $v_{\min} < p < v_{\max}$ such that $p \notin \mathrm{dom}(v)$ and $p - 1 \in \mathrm{dom}(v)$. Then $p > 2$. Let $w = (p-1)p \ (= \varrho_\kappa(M_{p-1}))$. Since $u$ contains $w$ or $\overline{w}$, it follows that either

$$u = u_{11}\overline{p}\,(\overline{p-1})v_1u_2, \text{ where } v = (\overline{p-1})v_1, \text{ or}$$
$$u = u_1v_1(p-1)pu_{21}, \text{ where } v = v_1(p-1).$$

(b) Similarly, if there exists $q$ with $v_{\min} < q < v_{\max}$ such that $q \notin \mathrm{dom}(v)$ and $q + 1 \in \mathrm{dom}(v)$, then either

$$u = u_{12}q(q+1)v_2u_2, \text{ where } v = (q+1)v_2, \text{ or}$$
$$u = u_1v_2\,(\overline{q+1})\overline{q}u_{22}, \text{ where } v = v_2(\overline{q+1}).$$

The following three cases can occur now and for each of them the conclusion from the statement of the lemma holds.

1. There are no pointers $p$ as in (a). In this case, $\mathrm{dom}(v) = [q+1, \kappa]$, where $q > 2$ is a unique pointer for which (b) holds. Also, if $\psi : \Delta_\kappa \to \Delta_\kappa$ is the morphism defined by $\psi(q+1) = 2$, $\psi(i) = i$ for any $2 \leq i \leq \kappa$, $i \neq q+1$, then $\psi(v)$ is realistic.
2. There are no pointers $q$ as in (b). In this case, $\mathrm{dom}(v) = [2, p-1]$, where $p \leq \kappa$ is a unique pointer for which (a) holds. Also, if $\psi : \Delta_\kappa \to \Delta_\kappa$ is the morphism defined by $\psi(p-1) = \kappa$, $\psi(i) = i$ for any $2 \leq i \leq \kappa$, $i \neq p-1$, then $\psi(v)$ is realistic.
3. There exists a pointer $p$ as in (a) and a pointer $q$ as in (b). By the above, $p$ is unique for (a) and $q$ is unique for (b). Hence either $\mathrm{dom}(v) = [q+1, p-1]$, if $p - 1 \geq q + 1$ or $\mathrm{dom}(v) = [2, p-1] \cup [q+1, \kappa]$, if $p - 1 < q + 1$.
   In the former case, we consider the morphism $\psi$ such that $\psi(p-1) = \kappa$, $\psi(q+1) = 2$, and $\psi(i) = i$, for any $2 \leq i \leq \kappa$, $i \neq q+1, p-1$. Then $\psi(v)$ is realizable.
   In the latter case, either $v = (q+1)w(p-1)$, in which case its conjugate $(p-1)(q+1)w$ is realistic, or $v = (\overline{p-1})w(\overline{q+1})$, in which case its conjugate $(\overline{q+1})(\overline{p-1})w$ is realistic. $\qquad\square$

*Example 10.8.* Let
$$u = \overline{7}\,\overline{6}\,7\,8\,2\,\overline{9}\,8\,\overline{3}\,\overline{2}\,9\,3\,4\,\overline{5}\,\overline{4}\,5\,6.$$

Then $u$ is realistic, since $u = \varrho_9(\overline{M}_6M_7M_1\overline{M}_8\overline{M}_2M_9M_3\overline{M}_4M_5)$. The string $u$ has the following legal proper substrings: (1) $v_1 = 4\,\overline{5}\,\overline{4}\,5$ with $\mathrm{dom}(v_1) = [4,5]$, (2) $v_2 = 8\,2\,\overline{9}\,8\,\overline{3}\,\overline{2}\,9\,3$ with $\mathrm{dom}(v_2) = [2,3] \cup [8,9]$, and (3) $v_3 = 8\,2\,\overline{9}\,8\,\overline{3}\,\overline{2}\,9\,3\,4\,\overline{5}\,\overline{4}\,5$ with $\mathrm{dom}(v_3) = [2,5] \cup [8,9]$. $\qquad\square$

The following result provides more details of the structure of legal substrings of realistic legal strings.

**Lemma 10.3.** *Let $u$ be a realistic string with $\mathrm{dom}(u) = [2, \kappa]$, and let $v$ be a legal substring of $u$.*

*(i) If $2 \notin \mathrm{dom}(v)$, then either $v = v_{\min}v'$ or $v = v'\overline{v}_{\min}$.*
*(ii) If $\kappa \notin \mathrm{dom}(v)$, then either $v = \overline{v}_{\max}v'$ or $v = v'v_{\max}$.*
*(iii) If $2, \kappa \in \mathrm{dom}(v)$, then $\mathrm{dom}(v) = [2, p] \cup [q, \kappa]$ and either $v = qv'p$ or $v = \overline{p}v'\overline{q}$.*

*Proof.* Assume that $v$ is a proper substring of $u$, that is, $v \neq u$.

Suppose first that $2 \notin \mathrm{dom}(v)$, and let $p = v_{\min}$. By Lemma 10.2, we have that $\mathrm{dom}(v) = [p, q]$ for some pointer $q$ with $p \leq q \leq \kappa$. Now since $p - 1 \notin \mathrm{dom}(v)$ and either $(p-1)p$ or $\overline{p}(\overline{p-1})$ is a substring of $u$, it must hold that either $v$ begins or $v$ ends with an occurrence of $p$. But this is possible only when $v = pv'$ or $v = v'\overline{p}$, as required.

The claim for the case $\kappa \notin \mathrm{dom}(v)$ is similar to the above case.

If $2, \kappa \in \mathrm{dom}(v)$, then $\mathrm{dom}(v) = [2, p] \cup [q, \kappa]$ for some $p$ and $j$ with $p < q$ by Lemma 10.2. Since either $p(p + 1)$ or $\overline{p}(\overline{p-1})$ is a substring of $u$, either $v = v_1p$ or $v = \overline{p}v_1$. Similarly, either $v = qv_2$ or $v = v_2\overline{q}$, and so the last claim from the statement of the lemma holds.    □

The following theorem gives a characterization of pairs of realistic legal strings that yield the same overlap graph.

Recall that for a string $v = p_1p_2\ldots p_n \in \Delta_\kappa^{\maltese}$, the reversal of $v$ is defined by $v^R = p_np_{n-1}\ldots p_1$, and the complementation by $v^C = \overline{p}_1\overline{p}_2\ldots\overline{p}_n$. For two signed strings $u, v \in \Delta_\kappa^{\maltese}$, let $u \approx v$ if $u$ is obtained from a conjugate of $v$ by a composition of the operations of reversal and complementation. Thus, $u \approx v$ if and only if there exist strings $v_1$ and $v_2$ such that $v = v_1v_2$ and $u$ is one of the strings $v_2v_1$, $(v_2v_1)^R$, $(v_2v_1)^C$, or $((v_2v_1)^R)^C$.

**Theorem 10.2.** *Let $u$ and $v$ be two realistic legal strings such that $\mathrm{dom}(u) = [2, \kappa] = \mathrm{dom}(v)$. Then $\gamma_u = \gamma_v$ if and only if $u \approx v$.*

*Proof.* Assume that $\gamma_u = \gamma_v$. We prove the theorem by induction on $\kappa$. If $\kappa = 2$, then clearly $u \approx v$. Suppose then that the claim holds for strings over $\Delta_{\kappa-1}$.

Since we now consider strings up to equivalence with respect to conjugation, reversal, and complementation, we can assume without loss of generality that both $u$ and $v$ end with $\varrho_\kappa(M_{k-1}) = (\kappa - 1)\kappa$. Now

$$u = u_1 \kappa' u_2(\kappa - 1)\kappa, \quad \text{and} \quad v = v_1 \kappa' v_2(\kappa - 1)\kappa, \tag{10.4}$$

where $\kappa' = \kappa$ or $\kappa' = \overline{\kappa}$. The strings $u' = u_1u_2(\kappa - 1)$ and $v' = v_1v_2(\kappa - 1)$, which are obtained from $u$ and $v$ by erasing the occurrences of $\kappa$, are realistic, and since $\gamma_{u'} = \gamma_{v'}$, we have $u' \approx v'$ by the induction hypothesis. Since the sign of the last occurrence of $(\kappa - 1)$ is the same in $u'$ and $v'$, complementation is not used in the above equivalence $u' \approx v'$. Therefore either $u'$ is a conjugate of $v'$ or of $(v')^R$. Moreover, the last occurrences $(\kappa - 1)$ in $u'$ and in $v'$ both

correspond to the end marker, which implies that $u' = v'$. If $u_2 = v_2$ then
also $u = v$, and the claim holds. Suppose then that $u \neq v$. Now either $u_2$
is a proper suffix of $v_2$ or $v_2$ is a proper suffix of $u_2$. By symmetry, we can
assume that the first alternative holds: $v_2 = wu_2$, and then also $u_1 = v_1 w$.
Since $\gamma_u = \gamma_v$, the substring $w$ must be a legal string (if only one occurrence
of a pointer $p$ is in $w$, then $\kappa$ would overlap with $p$ either in $u$ or $v$, but not
in both). Now

$$u = v_1 w \, \kappa' u_2 (\kappa - 1)\kappa, \quad \text{and} \quad v = v_1 \, \kappa' wu_2 (\kappa - 1)\kappa.$$

By Lemma 10.2, we have that $\mathrm{dom}(w) = [p, q]$ for some pointers $p \leq q$, and
then by Lemma 10.3(ii), either $w = w'q$ or $w = \overline{q}w'$ for a substring $w'$.
Moreover, as shown in the proof of Lemma 10.3, in the former case $w(q + 1)$
is a substring of both $u$ and $v$, and in the latter case $\overline{(q + 1)}w$ is a substring
of both $u$ and $v$. It then follows that $q = \kappa - 1$, and so $\varrho_\kappa(M_{\kappa-1})$ occurs twice
in $u$ and $v$. This is impossible since $u$ and $v$ are realistic.    □

## Notes on References

- Circle graphs (that is, overlap graphs of double occurrence strings) are also
  known as *interlacement graphs* (of self-intersecting curves). The interlacement graphs are well known in the connection of the Gauss Problem (see
  de Fraysseix and Ossona de Mendez [14]). For further properties of these
  graphs from the graph theoretic point of view, see, e.g., Bouchet [5, 6])
  and de Fraysseix [12].
- The following characterization of the overlap graphs is from Bouchet [6]: a
  simple graph $\gamma$ is an overlap graph of a double occurrence string if and only
  if no graph locally equivalent to $\gamma$ has an induced subgraph isomorphic to
  $W_5$ or $W_7$ or $BW_3$ of Fig. 10.8.

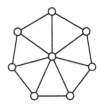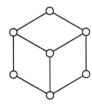

**Fig. 10.8.** The graphs $W_5$, $W_7$ and $BW_3$ in the characterization result of the overlap
graphs

- The results of Sect. 10.2 on realizable graphs were proved by Harju and
  Rozenberg [26]. The characterization result for the equivalence of overlap
  graphs in Sect. 10.3 is from Harju et al. [25].

# 11

# Graph Pointer Reduction System

In Chap. 10 we represented MDS structure of micronuclear and intermediate genes as graphs. In this chapter we continue this line of formalization by representing the gene assembly process in the framework of graphs. Thus, we shall formalize now the three molecular operations *ld*, *hi*, and *dlad* as operations (rewriting rules) on overlap graphs, obtaining in this way graph pointer reduction systems. Although the transition from legal strings to overlap graphs represents a considerable increase in abstraction (indeed, the linear structure of legal strings is lost!), we prove that the graph pointer reduction system is equivalent to the string pointer reduction system as far as strategies for gene assembly are concerned.

## 11.1 Assembly Operations on Graphs

We first define the graph pointer reduction system, and then we relate it to the string pointer reduction system on legal strings. We apply this system mostly to overlap graphs (of legal strings), but we begin with general signed graphs, for which we prove that the graph pointer reduction system is universal.

Let $S \subseteq V$ be a subset of the vertices of a signed graph $\gamma = (V, E, \sigma)$, where $\sigma \colon V \to \{-, +\}$ is the labelling function. The graph $\gamma' = (V, E', \sigma')$ is obtained from $\gamma$ by **complementing** $S$, if $\gamma'$ equals $\gamma$ with the subgraph induced by $S$ replaced by its complement (including the signs of the vertices in $S$). To be more precise, for all pairs $\{x, y\}$ of $V$ with $x \neq y$, let $\{x, y\} \in E'$ if and only if

$$\{x, y\} \in E, \quad \text{and} \quad x \notin S \text{ or } y \notin S, \text{ or}$$
$$\{x, y\} \notin E, \quad \text{and} \quad x \in S \text{ and } y \in S,$$

and $\sigma'(p) = -\sigma(p)$ for all $p \in S$, and $\sigma'(p) = \sigma(p)$ for all $p \notin S$. Moreover, if for a vertex $p \in V$, we complement the neighborhood $N_\gamma(p)$ of $p$, then we get the (signed) **local complement** $\mathrm{loc}_p(\gamma)$ at $p$ (in $\gamma$).

*Example 11.1.* Let the signed graph $\gamma$ be as in Fig. 11.1a. Then the graph that is obtained from $\gamma$ by complementing the set $S$ of shaded vertices is given in Fig. 11.1b.                                                                                $\square$

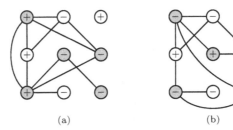

(a)                                                    (b)

**Fig. 11.1. a** Signed graph $\gamma$ from Example 11.1 where the shaded vertices form a chosen set $S$. **b** The graph where the subgraph induced by $S$ is complemented

Let $\gamma$ be a signed graph.

- The **graph negative rule** for a vertex $p$ is applicable to $\gamma$ if $p$ is isolated and negative in $\gamma$. The result $\mathsf{gnr}_p(\gamma)$ is the signed graph $\mathsf{gnr}_p(\gamma)$ obtained from $\gamma$ by removing the vertex $p$.
  Let
  $$\mathsf{Gnr} = \{\mathsf{gnr}_\mathbf{p} \mid p, q \in \Pi_\kappa,\ \kappa \geq 2\}$$
  be the set of all graph negative rules on overlap graphs of legal strings.

- The **graph positive rule** for a vertex $p$ is applicable to $\gamma$ if $p$ is positive in $\gamma$. The result $\mathsf{gpr}_p(\gamma)$ is the signed graph $\mathsf{gpr}_p(\gamma)$ obtained from the local complement $\mathsf{loc}_p(\gamma)$ by removing the vertex $p$.
  Let
  $$\mathsf{Gpr} = \{\mathsf{gpr}_\mathbf{p} \mid p \in \Pi_\kappa,\ \kappa \geq 2\}$$
  be the set of all graph positive rules on overlap graphs of legal strings.

- The **graph double rule** for different vertices $p$ and $q$ is applicable to $\gamma$ if $p$ and $q$ are adjacent and negative in $\gamma$. The result $\mathsf{gdr}_{p,q}(\gamma)$ is the signed graph, where $\mathsf{gdr}_{p,q}(\gamma) = (V \setminus \{p, q\}, E', \sigma')$ is obtained as follows: $\sigma'$ equals $\sigma$ restricted to $V \setminus \{p, q\}$, and $E'$ is obtained from $E$ by complementing the edges that join vertices in $N_\gamma(p)$ to vertices in $N_\gamma(q)$. This means that the status of a pair $\{x, y\}$ (for $x, y \in V \setminus \{p, q\}$) as an edge will change if and only if

$$x \in N_\gamma(p) \setminus N_\gamma(q), \quad \text{and} \quad y \in N_\gamma(q),$$
$$x \in N_\gamma(p) \cap N_\gamma(q), \quad \text{and} \quad y \in N_\gamma(q) \oplus N_\gamma(p),$$
$$x \in N_\gamma(q) \setminus N_\gamma(p), \quad \text{and} \quad y \in N_\gamma(p),$$

where $\oplus$ denotes the symmetric difference of the two neighborhoods, i.e., $(N_\gamma(q) \setminus N_\gamma(p)) \cup (N_\gamma(p) \setminus N_\gamma(q))$.
Let

$$\mathrm{Gdr} = \{\mathrm{gdr}_{p,q} \mid p, q \in \Pi_\kappa, \ \kappa \geq 2\}$$

be the set of all graph double rules on overlap graphs of legal strings.

*Example 11.2.* Consider the legal string $w = 3\bar{5}2\,6\,5\,4\,7\,3\,6\,7\,2\,\bar{4}$. Then $\gamma = \gamma_w$ is given in Fig. 11.2a, where the signs of the vertices are represented by superscripts. Recall that the vertices of $\gamma_w$ are the pointer sets $\mathbf{p} = \{p, \bar{p}\}$ for $p \in [2, 7]$. The operation $\mathrm{gpr}_4$ is applicable to $\gamma$, since the vertex $4$ is positive in the graph. The neighborhood of $4$ is $N_\gamma(4) = \{2, 3, 6\}$. The graph $\mathrm{gpr}_4(\gamma)$ is given in Fig. 11.2b.

Also, $\mathrm{gdr}_{2,3}$ is applicable to $\gamma$, since these vertices are negative and adjacent. Now $N_\gamma(2) = \{3, 4, 5, 6\}$, and $N_\gamma(3) = \{2, 4, 6, 7\}$. Therefore $N_\gamma(2) \cap N_\gamma(3) = \{4\}$, $N_\gamma(2) \setminus N_\gamma(3) = \{5\}$, and $N_\gamma(3) \setminus N_\gamma(2) = \{6, 7\}$. Then the graph $\mathrm{gdr}_{2,3}(\gamma)$ is given in Fig. 11.2c. $\qquad\square$

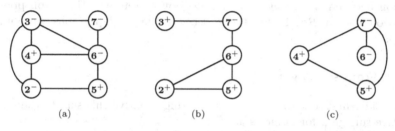

(a)                    (b)                    (c)

**Fig. 11.2. a** The graph $\gamma = \gamma_w$ from Example 11.2 for $w = 3\bar{5}2\,6\,5\,4\,7\,3\,6\,7\,2\,\bar{4}$. **b** The result of applying the rule $\mathrm{gpr}_4$ to $\gamma$. **c** The result of applying the rule $\mathrm{gdr}_{2,3}$ to $\gamma$

Let $\gamma$ be a signed graph on the set $V$ of vertices. A composition $\varphi = \varphi_n \dots \varphi_1$ of operations $\mathrm{gnr}_p$, $\mathrm{gpr}_p$ and $\mathrm{gdr}_{p,q}$, for $p, q \in V$, is called a **graph reduction** for $\gamma$ if $\varphi$ is applicable to $\gamma$, and $\varphi$ is **successful** for $\gamma$ if $\varphi(\gamma)$ is the empty graph (that has no vertices).

*Example 11.3.* The overlap graph $\gamma = \gamma_w$ given in Fig. 11.2a is reduced to the empty graph by the composition $\mathrm{gpr}_5 \, \mathrm{gpr}_6 \, \mathrm{gpr}_7 \, \mathrm{gpr}_4 \, \mathrm{gdr}_{2,3}$. Indeed, $\gamma_1 = \mathrm{gdr}_{2,3}(\gamma)$ is given in Fig. 11.2c, and $\gamma_2 = \mathrm{gpr}_4(\gamma_1)$ has a negative vertex $5$, positive vertices $6, 7$, and a single edge $\{5, 6\}$. Next $\gamma_3 = \mathrm{gpr}_7(\gamma_2)$ is $\gamma_2$ with the isolated positive vertex $7$ removed. Then $\mathrm{gpr}_6(\gamma_3)$ is the graph with a single (positive) vertex $5$, which is finally removed by $\mathrm{gnr}_5$. $\qquad\square$

The above operations are universal for signed graphs:

**Theorem 11.1.** *Let $\gamma$ be any signed graph. Then there exists a successful graph reduction $\varphi$ for $\gamma$.*

*Proof.* We prove the claim by contradiction. Assume then that $\gamma$ is a minimal counterexample to the claim, that is, $\gamma$ has the smallest number of vertices such that none of the operations are applicable for $\gamma$.

Now $\gamma$ has no positive vertices $p$ (for otherwise $\mathsf{gpr}_p$ would be applicable). Since no $\mathsf{gdr}_{p,q}$ can be applied, $\gamma$ has no edges, and hence it must have an isolated vertex $p$. But then $\mathsf{gnr}_p$ can be applied to $\gamma$ giving a smaller counterexample $\mathsf{gnr}_p(\gamma)$.                    □

## 11.2 Equivalence to String Pointer Reduction System

The **graph pointer reduction system** consists of the above graph reduction rules Gnr $\cup$ Gpr $\cup$ Gdr for overlap graphs on legal strings. We study now the relation between the string pointer reduction system and the graph pointer reduction system. Recall that the vertices of the overlap graphs are pointer sets **p**.

### 11.2.1 From snr to gnr

We consider first how to transform the string negative rule $\mathsf{snr}_p$ to the graph negative rule $\mathsf{gnr}_\mathbf{p}$ for pointers $p$.

**Lemma 11.1.** *Let $w \in \Delta_\kappa^{\maltese}$ be a legal string and $p \in \Pi_\kappa$ a pointer. If $\mathsf{snr}_p$ is applicable to $w$, then $\mathsf{gnr}_\mathbf{p}$ is applicable to $\gamma_w$. In this case,*

$$\gamma_{\mathsf{snr}_p(w)} = \mathsf{gnr}_\mathbf{p}(\gamma_w).$$

*Proof.* Assume that $\mathsf{snr}_p$ is applicable to $w$. Then $p$ is negative in $w$, and so the vertex **p** is negative in the overlap graph $\gamma_w$. Moreover, since $p$ does not overlap with any other pointer in $w$, **p** is isolated in $\gamma_w$. Consequently, $\mathsf{gnr}_\mathbf{p}$ is applicable to $\gamma_w$.

To prove the second claim, we note that $\mathsf{snr}_p$ removes the two consecutive occurrences of $p$ in $w$, and the overlap relation of the remaining pointers in $w$ remains the same in $w' = \mathsf{snr}(w)$. On the other hand, $\mathsf{gnr}_\mathbf{p}$ only removes the isolated vertex **p** from $\gamma_w$, and $\gamma' = \mathsf{gnr}_\mathbf{p}(\gamma_w)$ is an induced subgraph of $\gamma$ (with the same signing). Consequently, $\gamma_{w'} = \gamma'$, as required.                    □

## 11.2.2 From spr to gpr

We turn then to the relation between the graph positive rule $\text{gpr}_{\mathbf{p}}$ and the string operation $\text{spr}_p$. For this correspondence, we need an auxiliary operation $\text{sir}_p$ on the legal strings. For a pointer $p \in \Pi_\kappa$, we define $\text{sir}_p$ by

$$\text{sir}_p(w) = w_1 p \overline{w}_2 p' w_3, \quad \text{for } w = w_1 p w_2 p' w_3,$$

where $w_1, w_2, w_3 \in \Delta_\kappa^{\maltese}$, and $p' \in \mathbf{p}$. Hence, applying $\text{sir}_p$ to $w$ yields a legal string obtained from $w$ by inverting the "open $p$-interval" $w_2$.

**Lemma 11.2.** *Let $w \in \Delta_\kappa^{\maltese}$ be a legal string and $p \in \Pi_\kappa$. If $\text{sir}_p$ is applicable to $w$, then $\text{loc}_{\mathbf{p}}$ is applicable to $\gamma_w$, and*

$$\gamma_{\text{sir}_p(w)} = \text{loc}_{\mathbf{p}}(\gamma_w).$$

*Moreover, if $\text{spr}_p$ is applicable to $w$, then $\text{gpr}_{\mathbf{p}}$ is applicable to $\gamma_w$. In this case,*

$$\gamma_{\text{spr}_p(w)} = \text{gpr}_{\mathbf{p}}(\gamma_w).$$

*Proof.* Let $w = w_1 p w_2 p' w_3$ with $w_1, w_2, w_3 \in \Delta_\kappa^{\maltese}$ and $p' \in \mathbf{p}$. Hence $\text{sir}_p(w) = w_1 p \overline{w}_2 p' w_3$. The status of a pair $\{\mathbf{r}, \mathbf{s}\}$ as an edge is changed in the transformation $\gamma_w \mapsto \gamma_{\text{sir}_p(w)}$ if and only if exactly one occurrence of $r$ and one occurrence of $s$ lies in $w_2$. This is equivalent to saying that $\mathbf{r}, \mathbf{s} \in N_{\gamma_w}(\mathbf{p})$. Also, it is clear that the sign of a vertex $\mathbf{r}$ changes if and only if the pointer $r$ has exactly one occurrence in $w_2$, or equivalently if $\mathbf{r} \in N_{\gamma_w}(\mathbf{p})$. Hence $\gamma_{\text{sir}_p(w)} = \text{loc}_{\mathbf{p}}(\gamma_w)$. The second claim follows from the first one. $\quad\square$

## 11.2.3 From sdr to gdr

We consider next the transformation the string double rule $\text{sdr}_{p,q}$ to the graph double rule $\text{gnr}_{\mathbf{p},\mathbf{q}}$ for pointers $p$ and $q$.

**Lemma 11.3.** *Let $w \in \Delta_\kappa^{\maltese}$ be a legal string and $p, q \in \Pi_\kappa$ be pointers such that $\mathbf{p} \neq \mathbf{q}$. If $\text{sdr}_{p,q}$ is applicable to $w$, then $\text{gdr}_{\mathbf{p},\mathbf{q}}$ is applicable to $\gamma_w$. In this case,*

$$\gamma_{\text{sdr}_{p,q}(w)} = \text{gdr}_{\mathbf{p},\mathbf{q}}(\gamma_w).$$

*Proof.* Suppose that $\text{sdr}_{p,q}$ is applicable to $w$. Then the vertices $\mathbf{p}$ and $\mathbf{q}$ are adjacent and negative in $\gamma_w$, and thus the graph $\text{gdr}_{\mathbf{p},\mathbf{q}}(\gamma_w)$ is well defined.

We shall decompose the operation $\text{sdr}$ as follows:

$$\text{sdr}_{p,q} = \text{spr}_p \, \text{spr}_q \, \text{sir}_p \, .$$

Indeed, let $w = w_1 p w_2 q w_3 p w_4 q w_5$ for the substrings $w_i$, $i \in [1, 5]$. Then

$$\mathsf{spr}_p\,\mathsf{spr}_q\,\mathsf{sir}_p(w) = \mathsf{spr}_p\,\mathsf{spr}_q(w_1 p\overline{w}_3\overline{q}w_2 pw_4 qw_5)$$
$$= \mathsf{spr}_p(w_1 p\overline{w}_3\overline{w}_4\overline{p}w_2 w_5)$$
$$= w_1 w_4 w_3 w_2 w_5 = \mathsf{sdr}_{p,q}(w),$$

where in the first line $q$ and in the second line $p$ become positive, and therefore the operations $\mathsf{spr}_q$ and $\mathsf{spr}_p$ apply there. It is clear that the signs of the other pointers do not change in the transformation $w \mapsto \mathsf{sdr}_{p,q}(w)$.

Consider the transformation $\gamma_w \mapsto \gamma_{\mathsf{sdr}_{p,q}(w)}$. We shall decompose it into three steps corresponding to the decomposition of $\mathsf{sdr}_{p,q}$ as in the above:

$$\gamma_w \mapsto \gamma_{\mathsf{sir}_p(w)}, \qquad\qquad\qquad (\text{step 1})$$

$$\gamma_{\mathsf{sir}_p(w)} \mapsto \gamma_{\mathsf{spr}_q(\mathsf{sir}_p(w))}, \qquad\qquad (\text{step 2})$$

$$\gamma_{\mathsf{spr}_q(\mathsf{sir}_p(w))} \mapsto \gamma_{\mathsf{sdr}_{p,q}(w)}. \qquad\qquad (\text{step 3})$$

In Step 1, the neighborhood $N_{\gamma_w}(\mathbf{p})$ is complemented, and since $\mathbf{q} \in N_{\gamma_w}(\mathbf{p})$, $\mathbf{q}$ becomes positive in $\gamma_{\mathsf{sir}_p(w)}$, and

$$N_{\gamma_{\mathsf{sir}_p(w)}}(\mathbf{q}) = N_{\gamma_w}(\mathbf{p}) \oplus (N_{\gamma_w}(\mathbf{q}) \cup \{\mathbf{q}\}).$$

In Step 2, the neighborhood $N_{\gamma_{\mathsf{sir}_p(w)}}(\mathbf{q})$ is complemented, and $\mathbf{q}$ is removed from the graph. Since $p \in N_{\gamma_{\mathsf{sir}_p(w)}}(\mathbf{q})$, $\mathbf{p}$ becomes positive in $\gamma_{\mathsf{spr}_q(\mathsf{sir}_p(w))}$, and

$$N_{\gamma_{\mathsf{spr}_q(\mathsf{sir}_p(w))}}(\mathbf{p}) = N_{\gamma_w}(\mathbf{q}) \setminus \{\mathbf{p}\}.$$

In Step 3, the neighborhood $N_{\gamma_{\mathsf{spr}_q(\mathsf{sir}_p(w))}}(\mathbf{p})$ is complemented, and the vertex $\mathbf{p}$ is removed.

The pairs inside the three sets $N_{\gamma_w}(\mathbf{p}) \setminus N_{\gamma_w}(\mathbf{q})$, $N_{\gamma_w}(\mathbf{p}) \cap N_{\gamma_w}(\mathbf{q})$, and $N_{\gamma_w}(\mathbf{q}) \setminus N_{\gamma_w}(\mathbf{p})$ are complemented twice in the above steps, and hence the status of these pairs as edges remains the same in $\gamma_{\mathsf{sdr}_{p,q}(w)}$ as in $\gamma_w$. On the other hand, the pairs between the above three sets are complemented once in the transformation, and therefore they are complemented in $\gamma_w \mapsto \gamma_{\mathsf{sdr}_{p,q}(w)}$. This proves the claim. $\qquad\square$

The following result follows from Lemmas 11.1, 11.2, and 11.3:

**Theorem 11.2.** *Let $w \in \Delta_\kappa^{\maltese}$ be a legal string. Each string reduction $\varphi = \varphi_n \ldots \varphi_1$ for $w$ translates into a graph reduction $\varphi' = \varphi'_n \ldots \varphi'_1$ for an overlap graph $\gamma_w$ by the translation:*

$$\mathsf{snr}_p \mapsto \mathsf{gnr}_{\mathbf{p}}, \quad \mathsf{spr}_p \mapsto \mathsf{gpr}_{\mathbf{p}}, \quad \mathsf{sdr}_{p,q} \mapsto \mathsf{gdr}_{\mathbf{p},\mathbf{q}}.$$

*Consequently, if $\varphi$ is successful for $w$, then $\varphi'$ is successful for $\gamma_w$.*

### 11.2.4 Reverse Implications

We move now to prove the reverse implication, namely that any successful graph reduction $\varphi$ for an overlap graph $\gamma_w$ has an equivalent successful string reduction $\varphi'$ for the legal string $w$.

The correspondence from the legal strings to the overlap graphs is not injective that is, the same overlap graph represents several legal strings, as we saw in Chap. 10. Indeed, the vertices $\mathbf{p}$ of an overlap graph $\gamma_w$ do not specify which signs, $p \in \Delta_\kappa$ or $p \in \overline{\Delta}_\kappa$, the pointers have in the string $w$. In the following we denote by $\mathbf{p}(w)$ the **first occurrence** of $p$ or $\overline{p}$ in $w$.

*Example 11.4.* For the legal string $w = 2\,3\,\overline{4}\,\overline{5}\,3\,5\,\overline{4}\,2$, we have $\mathbf{2}(w) = 2$, $\mathbf{3}(w) = 3$, $\mathbf{4}(w) = \overline{4}$, and $\mathbf{5}(w) = \overline{5}$.    □

We begin with the positive rules $\mathsf{spr}_p$ and $\mathsf{gpr_p}$ and leave the more problematic negative rules as the last case.

**Lemma 11.4.** *Let $w \in \Delta_\kappa^{\circledast}$ be a legal string, and $p \in \Pi_\kappa$ a pointer. If $\mathsf{gpr_p}$ is applicable to $\gamma_w$, then $\mathsf{spr}_{\mathbf{p}(w)}$ is applicable to $w$. In this case,*

$$\gamma_{\mathsf{spr}_{\mathbf{p}(w)}(w)} = \mathsf{gpr_p}(\gamma_w).$$

*Proof.* If $\mathsf{gpr_p}$ is applicable to $\mathbf{p}$, then $\mathbf{p}$ is a positive vertex in $\gamma_w$, and hence $\mathbf{p}(w)$ is a positive pointer in $w$. Thus $\mathsf{spr}_{\mathbf{p}(w)}$ is applicable to $w$. The second claim follows from Lemma 11.2.    □

An analogous result holds also for the double rules:

**Lemma 11.5.** *Let $w \in \Delta_\kappa^{\circledast}$ be a legal string, and let $p, q \in \Pi_\kappa$ be pointers such that $\mathbf{p} \neq \mathbf{q}$. If $\mathsf{gdr_{p,q}}$ is applicable to $\gamma_w$, then $\mathsf{sdr}_{\mathbf{p}(w),\mathbf{q}(w)}$ is applicable to $w$. In this case,*

$$\gamma_{\mathsf{sdr}_{\mathbf{p}(w),\mathbf{q}(w)}(w)} = \mathsf{gdr_{p,q}}(\gamma_w).$$

*Proof.* If $\mathsf{gdr_{p,q}}$ is applicable to $\gamma_w$, then the vertices $\mathbf{p}$ and $\mathbf{q}$ are both negative and they are adjacent in $\gamma_w$. Consequently, the pointers $\mathbf{p}(w)$ and $\mathbf{q}(w)$ are negative and they overlap in the legal string $w$. Hence $\mathsf{sdr}_{\mathbf{p}(w),\mathbf{q}(w)}$ is applicable to $w$. The second claim follows from Lemma 11.3.    □

There is no similar result for the negative rules. Indeed, for the legal string $w = pqqp$, the vertex $\mathbf{p}$ of $\gamma_w$ is negative and is isolated in $\gamma_w$. Hence $\mathsf{gnr_p}$ is applicable to $\gamma_w$, but, on the other hand, $\mathsf{snr}_p$ is not applicable to $w$. However, we show that one can *translate* successful graph reductions into successful string reductions. For this we need the notion of permutation of a composition.

We say that a composition $\varphi$ of string operations is **canonical** if it has the form $\varphi = \rho\omega$, where $\rho = \mathsf{snr}_{p_n}\,\mathsf{snr}_{p_{n-1}}\ldots\mathsf{snr}_{p_1}$ consists of string negative rules only, and the composition $\omega$ has no string negative rules. Similarly, a composition $\varphi$ of graph operations is **canonical** if it has the form $\varphi = \rho\omega$, where $\rho = \text{gnr}_{p_n} \, \text{gnr}_{p_{n-1}} \ldots \text{gnr}_{p_1}$ consists of graph negative rules only, and the composition $\omega$ has no graph negative rules.

A **canonical reduction** for graphs is defined similarly (the graph negative rules $\text{gnr}_{\mathbf{p}}$, resp.).

**Lemma 11.6.** *Let $w$ be a legal string, and let $\varphi = \varphi_n \ldots \varphi_1$ be a successful string reduction for $w$. Then there exists a permutation $(i_1 \ldots i_n)$ of $[1, n]$ such that $\varphi' = \varphi_{i_n} \ldots \varphi_{i_1}$ is a canonical string reduction successful for $w$.*

*Proof.* By Theorem 9.1, there exists a successful string reduction $\varphi$ for $w$. The claim of the lemma follows easily from the fact that removing a negative pointer from $w$ does not change the rest of the string. Indeed, if $\varphi_i = \text{snr}_p$ and $\varphi_{i+1} = \text{spr}_q$ or $\varphi_{i+1} = \text{sdr}_{q,r}$ in the composition $\varphi$, then applying $\varphi_{i+1}$ before $\varphi_i$ also yields a successful string reduction for $w$. □

The corresponding result for graphs is given in the following lemma.

**Lemma 11.7.** *Let $\gamma$ be a signed graph, and let $\varphi = \varphi_n \ldots \varphi_1$ be a successful graph reduction for $\gamma$. Then there exists a permutation $(i_1 \ldots i_n)$ of $[1, n]$ such that $\varphi' = \varphi_{i_n} \ldots \varphi_{i_1}$ is a canonical graph reduction successful for $\gamma$.*

*Proof.* The claim of the lemma follows easily from the fact that removing a negative vertex does not change anything in the rest of the graph. Indeed, if $\varphi_i = \text{gnr}_{\mathbf{p}}$ and $\varphi_{i+1} = \text{gpr}_{\mathbf{q}}$ or $\varphi_{i+1} = \text{gdr}_{\mathbf{q,r}}$ in the composition $\varphi$, then applying $\varphi_{i+1}$ before $\varphi_i$ also yields a successful graph reduction for $\gamma$. □

Two compositions $\rho = \rho_n \ldots \rho_1$ and $\rho' = \rho'_n \ldots \rho'_1$ of string operations (graph operations, resp.) are said to be **permutations** of each other if there is a permutation $(i_1 \ldots i_n)$ of $[1, n]$ such that $\rho'_j = \rho_{i_j}$.

**Lemma 11.8.** *Let $\varphi = \rho\omega$ be a canonical graph reduction, where $\rho$ consists of the graph negative rules in $\varphi$. If $\varphi$ is successful for $\gamma$, then so is $\varphi' = \rho'\omega$ for each permutation $\rho'$ of $\rho$.*

*Proof.* Since negative rules apply to isolated vertices only, any two consecutive applications of negative rules can be permuted without changing the result of the whole reduction. □

Note that Lemma 11.8 does not have a corresponding result for strings. Indeed, for $w = qppq$, the string reduction $\text{snr}_q \, \text{snr}_p$ is successful for $w$, while $\text{snr}_p \, \text{snr}_q$ is not.

**Lemma 11.9.** *Let $w$ be a legal string. If $\varphi = \text{gnr}_{\mathbf{p}_n} \ldots \text{gnr}_{\mathbf{p}_1}$ is successful for $\gamma_w$, then there exists a permutation $(i_1 \ldots i_n)$ of $[1, n]$ such that $\varphi' = \text{snr}_{\mathbf{p}_{i_n}(w)} \ldots \text{snr}_{\mathbf{p}_{i_1}(w)}$ is successful for $w$ and $\varphi'' = \text{gnr}_{\mathbf{p}_{i_n}} \ldots \text{gnr}_{\mathbf{p}_{i_1}}$ is successful for $\gamma_w$.*

*Proof.* We prove the claim by induction on the length $n$ of the compositions. If $n = 1$, then $w = pp$ for a pointer $p$, and thus $\varphi' = \mathsf{snr}_p$ is successful for $w$ and $\varphi'' = \mathsf{gnr}_{\mathbf{p}}$ is successful for $\gamma_w$.

Assume now that the claim holds for compositions of length at most $n - 1$, and let the length of $\varphi$ be $n$. Since $\varphi$ has only negative rules, $\gamma_w$ consists of negative isolated vertices only. Hence no two pointers overlap in $w$, and so the intervals of the pointers in $w$ represent a (nested) parenthesis structure. Therefore there exists a pointer $q = p_{i_1}$ such that $qq$ is a substring of $w$. Now the rule $\mathsf{snr}_q$ is applicable to $w$. Let $\varphi = \rho_2\, \mathsf{gnr}_{\mathbf{q}}\, \rho_1$, and let $\varphi_1 = \rho_2\rho_1\, \mathsf{gnr}_{\mathbf{q}}$. By Lemma 11.8, $\varphi_1$ is successful for $\gamma_w$. The composition $\rho_2\rho_1$ has length $n - 1$, and hence, by the induction hypothesis, the claim holds for $\rho_2\rho_1$ (and the graph $\mathsf{gnr}_{\mathbf{q}}$). Hence the claim holds also for $\varphi$. $\qquad\square$

We collect together the results of this section in the following theorem.

**Theorem 11.3.** *Let $w$ be a legal string, and let $\varphi$ be a successful graph reduction for $\gamma_w$. Then there exists a permutation $\varphi'' = \varphi_n \ldots \varphi_1$ of $\varphi$, which is successful graph reduction for $\gamma_w$, and which can be translated to a successful string reduction $\varphi' = \varphi'_n \ldots \varphi'_1$ for $w$ by the following translations:*

$$\mathsf{gnr}_{\mathbf{p}} \mapsto \mathsf{snr}_{\mathbf{p}(w)}, \quad \mathsf{gpr}_{\mathbf{p}} \mapsto \mathsf{spr}_{\mathbf{p}(w)}, \quad \mathsf{gdr}_{\mathbf{p},\mathbf{q}} \mapsto \mathsf{sdr}_{\mathbf{p}(w),\mathbf{q}(w)} \, .$$

*Proof.* By Lemma 11.7, we can assume that the composition $\varphi$ is canonical. Hence $\varphi = \mathsf{gnr}_{\mathbf{p}_j} \ldots \mathsf{gnr}_{\mathbf{p}_1} \rho_i \ldots \rho_1$ for some vertices $\mathbf{p}_1, \ldots, \mathbf{p}_j$ and some positive or double graph rules $\rho_1, \ldots, \rho_i$. For each $k$, let $\rho'_k$ be obtained from $\rho_k$ as follows:

$$\rho'_k = \begin{cases} \mathsf{spr}_{\mathbf{p}(w)}, & \text{if } \rho_k = \mathsf{gpr}_{\mathbf{p}}, \\ \mathsf{sdr}_{\mathbf{p}(w),\mathbf{q}(w)}, & \text{if } \rho_k = \mathsf{gdr}_{\mathbf{p},\mathbf{q}}, \end{cases}$$

for some pointers $p$ and $q$. By Lemmata 11.4 and 11.5, $\varphi_1 = \rho'_i, \ldots, \rho'_1$ is applicable to $\gamma_w$, and for $\gamma' = \rho_i \ldots \rho_1(\gamma_w)$ and $w' = \rho'_i \ldots \rho'_1(w)$, we have $\gamma' = \gamma_{w'}$. Then, by Lemma 11.9, there exists a successful string reduction $\mathsf{snr}_{q_j} \ldots \mathsf{snr}_{q_1}$ for $w'$, as required. This proves the theorem. $\qquad\square$

## Notes on References

- The graph pointer reduction system was introduced by Ehrenfeucht et al. [15, 20]. In the first of these articles the operations $\mathsf{gnr}$ and $\mathsf{gpr}$ were investigated and their correspondence with the string operations $\mathsf{snr}$ and $\mathsf{spr}$ were established. In [15] the remaining graph operation $\mathsf{gdr}$ was studied.
- Lemma 11.2 is considered by Bouchet [6] for circle graphs of double occurrence strings. In the present form stated in this chapter for signed overlap graphs, it is proved in [15].
- The proof of Lemma 11.3 is adapted from Bouchet [6], where a closely related result is stated for (unsigned) overlap graphs and double occurrence strings.

# Part III

Properties of Gene Assembly

# 12

# Invariants

Our model for the process of gene assembly in ciliates is highly nondeterministic in the sense that for a given micronuclear gene there may exist many different strategies for assembling that gene using the three operations $ld$, $hi$, and $dlad$. For instance, using computer simulations, we noticed that there are 3,060 different sequential strategies to assemble the actin gene in *Sterkiella nova*, and 23,940 different sequential assembly strategies for the same gene in *Sterkiella trifallax*. For this reason, it is important to study the *invariants* of gene assembly, i.e., those properties of gene assembly that are common to all these strategies.

The $ld$-rule yields a linear molecule by a simple $ld$-rule and a circular molecule by a boundary operation, and therefore a macronuclear gene can be assembled either in a linear or in a circular molecule — we prove that this is an invariant and so it does not depend on the chosen assembly strategy. We prove that the *context* of the gene, i.e., the exact sequence of IESs preceding and succeeding the assembled gene, is also an invariant of the process. Still another invariant of the assembly strategies is the set of "residual" molecules, i.e., all the other molecules produced in the process containing IES sequences only. This implies in particular that the number of $ld$ operations used in any assembly is also an invariant, although the set of pointers on which $ld$ is applied may vary from assembly to assembly.

## 12.1 MDS–IES Descriptors

The MDS descriptors as defined and treated in Chap. 6 give explicitly only the (pairs of pointers of) MDSs as sequences in $\Gamma_\kappa^\maltese$. In this chapter we shall extend this notation to MDS–IES descriptors in order to have also ("the trace of") the interspersing IESs to be represented in the sequences.

Tracing IESs (i.e., what happens to IESs) is of interest from the biological point of view. Moreover, from the point of view of our approach to the study of gene assembly, tracing IESs is also natural.

Let

$$\Omega = \{I_1, I_2, \dots\}$$

be the **alphabet of IESs**. A signed string $\iota$ over $\Omega$ is called an **IES sequence** if for any $I \in \Omega$, $\iota$ contains at most one element from $\{I, \bar{I}\}$.

A signed string $\delta$ over $\Gamma_\kappa \cup \Omega$ is called an **MDS–IES descriptor** if

$$\delta = \iota_1(p_1, q_1)\iota_2(p_2, q_2) \dots \iota_n(p_n, q_n)\iota_{n+1}, \qquad (12.1)$$

where $n \geq 1$, $\iota_j \in \Omega^{\maltese}$ for $j \in [1, n+1]$, and $(p_i, q_i) \in \Gamma_\kappa \cup \bar{\Gamma}_\kappa$ for each $i \in [1, n]$, and moreover, the following conditions are satisfied: for $\delta' = (p_1, q_1)(p_2, q_2) \dots (p_n, q_n)$ and $\delta'' = \iota_1\iota_2 \dots \iota_n\iota_{n+1}$,

(i) $\delta'$ is a realistic MDS descriptor,
(ii) $\delta''$ is an IES sequence,
(iii) $\iota_i \neq \Lambda$ for $2 \leq i \leq n$.

We call $\delta'$ the **MDS-trace**, and $\delta''$ is the **IES-trace** of $\delta$. We also say that $\delta$ is an **MDS–IES micronuclear descriptor** if

$$\delta = (p_1, q_1)I_1(p_2, q_2)I_2 \dots I_{n-1}(p_n, q_n),$$

where the MDS-trace of $\delta$ is a micronuclear descriptor, as defined in Chap. 6.

*Example 12.1.* Let $\kappa = 11$. Then $\delta = \bar{I}_7(\bar{5}, \bar{2})I_8 I_3 I_{10}(b, 2)\bar{I}_9(11, e)I_4(5, 11)$ is an MDS–IES descriptor, where the MDS-trace is $\delta' = (\bar{5}, \bar{2})(b, 2)(11, e)(5, 11)$, and the IES-trace of $\delta$ is $\delta'' = \bar{I}_7 I_8 I_3 I_{10} \bar{I}_9 I_4$.  □

For an MDS–IES descriptor $\delta$ and an IES sequence $\iota$, we say that $\iota$ is **left** (**right**, resp.) **bordered** by $p$ in $\delta$ if there is a pointer or marker $q$ in $\delta$ such that $(q, p)\iota$ is a substring ($\iota(p, q)$ is a substring, resp.) of either $\delta$ or the inversion $\bar{\delta}$.

The molecular gene assembly operations can be formalized as in Chap. 7, but now for MDS–IES descriptors. We shall use the notations $\underline{\mathsf{ld}}$, $\underline{\mathsf{hi}}$, and $\underline{\mathsf{dlad}}$ for the three operations in this expanded formalism corresponding to ld, hi, and dlad.

In what follows, for an operation $\varphi$ and an MDS–IES descriptor $\delta$, we shall write $\varphi(\delta) = \delta_1 + \delta_2$ to denote the fact that $\delta$ produces two MDS–IES descriptors $\delta_1$ and $\delta_2$ from the given MDS–IES descriptor $\delta$ (as will be the case for the $\underline{\mathsf{ld}}$-rule). Also, each operation $\varphi$ is extended to sets of linear and circular strings (in the sum formalism) by setting

$$\varphi(\delta_1 + \delta_2 + \dots + \delta_t) = \varphi(\delta_1) + \varphi(\delta_2) + \dots + \varphi(\delta_t),$$

and also

$$\varphi([\delta]) = [\varphi(\delta)],$$

for all MDS–IES descriptors $\delta_i$, $i = 1, 2, \dots, t$, where $[v]$ denotes the circular string of the string $v$, see Sect. 5.4. We have naturally that $\delta_1 + \delta_2 = \delta_2 + \delta_1$ and $\delta + \Lambda = \delta$ for all MDS–IES descriptors $\delta_1$, $\delta_2$, and $\delta$.

We are now ready to introduce the **MDS–IES pointer reduction system.**

(1) For each $p \in \Pi_\kappa$, the $\underline{\text{ld}}$-rule for $p$ is defined by

$$\underline{\text{ld}}_p(\delta_1(q,p)\iota_1(p,r)\delta_2) = \delta_1(q,r)\delta_2 + [\iota_1], \qquad (\ell a1)$$

$$\underline{\text{ld}}_p(\iota_1(p,m)\iota_2(m',p)\iota_3) = \iota_1\iota_3 + [(m',m)\iota_2], \qquad (\ell a2')$$

where $q, r \in \Pi_\kappa$, $\delta_1, \delta_2 \in (\Gamma_\kappa \cup \Omega)^{\maltese}$, $\iota_1, \iota_2, \iota_3 \in \Omega^{\maltese}$, and $m, m' \in \mathfrak{M}$.

We say that the circular string $[\iota_1]$ (the signed string $\iota_1\iota_3$, resp.) is **excised** from $\delta$ in Case ($\ell a1$) (in Case ($\ell a2'$), resp.).

Case ($\ell a1$) is called a **simple ld-rule**, while Case ($\ell a2'$) is called a **boundary ld-rule**, cf. Sect. 7.1.

We notice that the circular strings are now explicitly present in both of the defined equations (this was *not* the case for the MDS-descriptors – see ($\ell 1$) on p. 76 and ($\ell 2'$) on p. 79). Also, the IES-trace is either shorter in the resulting MDS–IES descriptor $\delta_1(q,r)\delta_2$, or the IES sequence $\iota_2$ present in the circular string $[(m',m)\iota_2]$ is not longer than the original IES-trace (recall that in Chap. 7 we adopted the convention that the boundary $\underline{\text{ld}}$-rule is applied only as the last operation).

In the case of the $\underline{\text{hi}}$-rule and the $\underline{\text{dlad}}$-rule the only change in comparison to the definitions in Chap. 7 is the requirement that we now have $\delta_i \in (\Gamma_\kappa \cup \Omega)^{\maltese}$ rather than $\delta_i \in \Gamma_\kappa^{\maltese}$. This is because only the $\underline{\text{ld}}$-rule splits a (linear) string into a (linear) string and a circular string.

(2) For each $p \in \Pi_\kappa$, the $\underline{\text{hi}}$-rule for $p$ is defined by

$$\underline{\text{hi}}_p(\delta_1(p,q)\delta_2(\overline{p},\overline{r})\delta_3) = \delta_1\overline{\delta}_2(\overline{q},\overline{r})\delta_3,$$
$$\underline{\text{hi}}_p(\delta_1(q,p)\delta_2(\overline{r},\overline{p})\delta_3) = \delta_1(q,r)\overline{\delta}_2\delta_3,$$

where $q, r \in \Pi_\kappa$, and $\delta_1, \delta_2 \in (\Gamma_\kappa \cup \Omega)^{\maltese}$.

(3) For each $p, q \in \Pi_\kappa$, $p \neq q$, the $\underline{\text{dlad}}$-rule for $p$ and $q$ is defined by

$$\underline{\text{dlad}}_{p,q}(\delta_1(p,r_1)\delta_2(q,r_2)\delta_3(r_3,p)\delta_4(r_4,q)\delta_5) = \delta_1\delta_4(r_4,r_2)\delta_3(r_3,r_1)\delta_2\delta_5,$$
$$\underline{\text{dlad}}_{p,q}(\delta_1(p,r_1)\delta_2(r_2,q)\delta_3(r_3,p)\delta_4(q,r_4)\delta_5) = \delta_1\delta_4\delta_3(r_3,r_1)\delta_2(r_2,r_4)\delta_5,$$
$$\underline{\text{dlad}}_{p,q}(\delta_1(r_1,p)\delta_2(q,r_2)\delta_3(p,r_3)\delta_4(r_4,q)\delta_5) = \delta_1(r_1,r_3)\delta_4(r_4,r_2)\delta_3\delta_2\delta_5,$$
$$\underline{\text{dlad}}_{p,q}(\delta_1(r_1,p)\delta_2(r_2,q)\delta_3(p,r_3)\delta_4(q,r_4)\delta_5) = \delta_1(r_1,r_3)\delta_4\delta_3\delta_2(r_2,r_4)\delta_5,$$
$$\underline{\text{dlad}}_{p,q}(\delta_1(p,r_1)\delta_2(q,p)\delta_4(r_4,q)\delta_5) = \delta_1\delta_4(r_4,r_1)\delta_2\delta_5,$$
$$\underline{\text{dlad}}_{p,q}(\delta_1(p,q)\delta_3(r_3,p)\delta_4(q,r_4)\delta_5) = \delta_1\delta_4\delta_3(r_3,r_4)\delta_5,$$
$$\underline{\text{dlad}}_{p,q}(\delta_1(r_1,p)\delta_2(q,r_2)\delta_3(p,q)\delta_5) = \delta_1(r_1,r_2)\delta_3\delta_2\delta_5,$$

where $r_1, r_2, r_3, r_4, r_5 \in \Pi_\kappa$, and $\delta_1, \delta_2, \delta_3, \delta_4, \delta_5 \in (\Gamma_\kappa \cup \Omega)^{\maltese}$.

For all $\varphi \in \{\underline{\text{ld}}_p, \underline{\text{hi}}_p, \underline{\text{dlad}}_{p,q} \mid p, q \in \Pi_\kappa, \kappa \geq 2\}$ and IES sequences $\iota$, we set $\varphi(\iota) = \iota$. Therefore, we have the following simple result.

**Lemma 12.1.** *Let $\delta$ be an MDS–IES descriptor and $\varphi$ a composition of operations from $\{\underline{\text{ld}}_p, \underline{\text{hi}}_p, \underline{\text{dlad}}_{p,q} \mid p, q \in \Pi_\kappa, \kappa \geq 2\}$. Then*

*(1) $\varphi(\delta)$ consists of one linear string and possibly several circular strings*

*(2) $\varphi(\delta)$ contains one MDS–IES descriptor, and one or more IES sequences.*

Note, however, that the MDS–IES component of $\varphi(\delta)$ in the above lemma may not coincide with the linear component of $\varphi(\delta)$. In fact, they coincide if and only if $\varphi$ contains no boundary $\underline{\text{ld}}$ operations. Note also that the linear string of $\varphi(\delta)$ may be the empty string — this happens if and only if $\delta$ starts and ends with a same pointer.

We shall now modify some terminology concerning MDS descriptors in the case of MDS–IES descriptors. An MDS–IES descriptor $\delta$ is **assembled** if there are IES sequences $\iota_1, \iota_2$ such that

(i) $\delta = \iota_1(b, e)\iota_2$ – we say then that the descriptor has been assembled **linearly** in the **orthodox order**, or

(ii) $\delta = \iota_1(\overline{e}, \overline{b})\iota_2$ – we say then that the descriptor has been assembled **linearly** in the **inverted order**, or

(iii) $\delta = [(b, e)\iota_1]$ – we say then that the descriptor has been assembled **circularly**.

As before, we say that a composition $\varphi$ of operations from $\{\underline{\text{ld}}_p, \underline{\text{hi}}_p, \underline{\text{dlad}}_{p,q} \mid p, q \in \Pi_\kappa, \kappa \geq 2\}$ is an **assembly strategy** for $\delta$ if $\varphi$ is applicable to $\delta$. Now, we say that an assembly strategy $\varphi$ is **successful** for $\delta$ if $\varphi(\delta)$ contains an assembled MDS–IES descriptor – we also say that $\varphi(\delta)$ is an **assembled set** of $\delta$.

By the proof of Theorem 7.1, if an MDS–IES descriptor $\delta$ has an occurrence of a pair $(p, q)$, where either $p$ or $q$ is not a marker, then there exists an operation that is applicable to $\delta$. Hence, the following result is a corollary of the proof of Theorem 7.1 and Lemma 12.1.

**Theorem 12.1.** *Every MDS–IES descriptor $\delta$ has a successful assembly strategy $\varphi$. Moreover, there is an integer $t \geq 2$ and IES sequences $\iota_i$ such that either*

$$\varphi(\delta) = \iota_1(b, e)\iota_2 + [\iota_3] + \cdots + [\iota_t], \tag{12.2}$$

*or*

$$\varphi(\delta) = \iota_1(\overline{e}, \overline{b})\iota_2 + [\iota_3] + \cdots + [\iota_t], \tag{12.3}$$

*or*

$$\varphi(\delta) = [(b, e)\iota_1] + \iota_2 + [\iota_3] + \cdots + [\iota_t]. \tag{12.4}$$

The **context** of $\varphi(\delta)$, denoted by $\text{con}(\varphi(\delta))$, is

- the pair $(\iota_1, \iota_2)$ in Cases (12.2) and (12.3), and

- the string $\iota_1$ in Case (12.4).

The string $\iota_2$ in (12.4) and the circular strings $[\iota_i]$, for $i \in [3, t]$, in (12.2), (12.3), and (12.4) are called the **residual strings** of $\varphi(\delta)$. The set of residual strings of $\varphi(\delta)$ is denoted by $\mathrm{res}(\varphi(\delta))$.

*Example 12.2.* Consider the MDS–IES descriptor

$$\delta = (\overline{10}, \overline{8})I_1(\overline{3}, \overline{b})I_2(\overline{5}, \overline{3})I_3(10, 11)I_4(5, 8)I_5(11, e).$$

As is mostly the case, $\delta$ does not have a unique successful assembly strategy. E.g., the compositions

$$\varphi_1 = \underline{\mathsf{dlad}}_{3,11}\,\underline{\mathsf{dlad}}_{5,8}\,\underline{\mathsf{hi}}_{\overline{10}} \quad \text{and} \quad \varphi_2 = \underline{\mathsf{hi}}_{\overline{10}}\,\underline{\mathsf{hi}}_{\overline{11}}\,\underline{\mathsf{hi}}_{\overline{3}}\,\underline{\mathsf{hi}}_{\overline{8}}\,\underline{\mathsf{hi}}_{\overline{5}}$$

are both successful as seen from the following:

$$\underline{\mathsf{hi}}_{\overline{10}}(\delta) = \overline{I}_3(3,5)\overline{I}_2(b,3)\overline{I}_1(8,11)I_4(5,8)I_5(11,e),$$
$$\underline{\mathsf{dlad}}_{5,8}(\underline{\mathsf{hi}}_{\overline{10}}(\delta)) = \overline{I}_3(3,11)I_4\overline{I}_2(b,3)\overline{I}_1 I_5(11,e),$$
$$\underline{\mathsf{dlad}}_{3,11}(\underline{\mathsf{dlad}}_{5,8}(\underline{\mathsf{hi}}_{\overline{10}}(\delta))) = \overline{I}_3\overline{I}_1 I_5 I_4\overline{I}_2(b,e),$$

and

$$\underline{\mathsf{hi}}_{\overline{5}}(\delta) = (\overline{10}, \overline{8})I_1(\overline{3}, \overline{b})I_2\overline{I}_4(\overline{11}, \overline{10})\overline{I}_3(3,8)I_5(11,e),$$
$$\underline{\mathsf{hi}}_{\overline{8}}(\underline{\mathsf{hi}}_{\overline{5}}(\delta)) = (\overline{10}, \overline{3})I_3(10,11)I_4\overline{I}_2(b,3)\overline{I}_1 I_5(11,e),$$
$$\underline{\mathsf{hi}}_{\overline{3}}(\underline{\mathsf{hi}}_{\overline{8}}(\underline{\mathsf{hi}}_{\overline{5}}(\delta))) = (\overline{10}, \overline{b})I_2\overline{I}_4(\overline{11}, \overline{10})\overline{I}_3\overline{I}_1 I_5(11,e),$$
$$\underline{\mathsf{hi}}_{\overline{11}}(\underline{\mathsf{hi}}_{\overline{3}}(\underline{\mathsf{hi}}_{\overline{8}}(\underline{\mathsf{hi}}_{\overline{5}}(\delta)))) = (\overline{10}, \overline{b})I_2\overline{I}_4\overline{I}_5 I_1 I_3(10,e),$$
$$\underline{\mathsf{hi}}_{\overline{10}}(\underline{\mathsf{hi}}_{\overline{11}}(\underline{\mathsf{hi}}_{\overline{3}}(\underline{\mathsf{hi}}_{\overline{8}}(\underline{\mathsf{hi}}_{\overline{5}}(\delta))))) = \overline{I}_3\overline{I}_3 I_5 I_4\overline{I}_2(b,e).$$

We notice that $\mathrm{con}(\varphi_1) = \mathrm{con}(\varphi_2) = (\overline{I}_3\overline{I}_1 I_5 I_4\overline{I}_2, \Lambda)$, $\mathrm{res}(\varphi_1) = \mathrm{res}(\varphi_2) = \emptyset$, and the number of $\underline{\mathsf{ld}}$-rules in both $\varphi_1$ and $\varphi_2$ is equal to zero. □

*Example 12.3.* Consider the MDS descriptor

$$\delta = (8, 10)I_1(3, 4)I_2(b, 3)I_3(5, 8)I_4(12, e)I_5(4, 5)I_6(10, 12).$$

Here, e.g., both

$$\varphi_1 = \underline{\mathsf{ld}}_{12}\,\underline{\mathsf{ld}}_5\,\underline{\mathsf{dlad}}_{8,10}\,\underline{\mathsf{dlad}}_{3,4} \quad \text{and} \quad \varphi_2 = \underline{\mathsf{ld}}_8\,\underline{\mathsf{ld}}_{10}\,\underline{\mathsf{dlad}}_{3,4}\,\underline{\mathsf{dlad}}_{5,12}$$

are successful assembly strategies. In more detail, we have

$$\underline{\mathsf{dlad}}_{3,4}(\delta) = (8, 10)I_1 I_3(5, 8)I_4(12, e)I_5 I_2(b, 5)I_6(10, 12),$$
$$\underline{\mathsf{dlad}}_{8,10}(\underline{\mathsf{dlad}}_{3,4}(\delta)) = I_4(12, e)I_5 I_2(b, 5)I_6 I_1 I_3(5, 12),$$
$$\underline{\mathsf{ld}}_5(\underline{\mathsf{dlad}}_{8,10}(\underline{\mathsf{dlad}}_{3,4}(\delta))) = I_4(12, e)I_5 I_2(b, 12) + [I_6 I_1 I_3],$$
$$\underline{\mathsf{ld}}_{12}(\underline{\mathsf{ld}}_5(\underline{\mathsf{dlad}}_{8,10}(\underline{\mathsf{dlad}}_{3,4}(\delta)))) = [(b, e)I_5 I_2] + I_4 + [I_6 I_1 I_3],$$

and

$$\underline{\mathsf{dlad}}_{5,12}(\delta) = (8,10)I_1(3,4)I_2(b,3)I_3I_6(10,e)I_5(4,8)I_4,$$
$$\underline{\mathsf{dlad}}_{3,4}(\underline{\mathsf{dlad}}_{5,12}(\delta)) = (8,10)I_1I_3I_6(10,e)I_5I_2(b,8)I_4,$$
$$\underline{\mathsf{ld}}_{10}(\underline{\mathsf{dlad}}_{3,4}(\underline{\mathsf{dlad}}_{5,12}(\delta))) = (8,e)I_5I_2(b,8)I_4 + [I_1I_3I_6],$$
$$\underline{\mathsf{ld}}_8(\underline{\mathsf{ld}}_{10}(\underline{\mathsf{dlad}}_{3,4}(\underline{\mathsf{dlad}}_{5,12}(\delta)))) = [(b,e)I_5I_2] + I_4 + [I_1I_3I_6].$$

In both strategies, the string $(b,e)I_5I_2$ is obtained by a boundary $\underline{\mathsf{ld}}$-rule, and therefore it represents a circular molecule, where the ends of the gene are joined by the composite IES string $I_5I_2$. Now, $\mathrm{con}(\varphi_1) = \mathrm{con}(\varphi_2) = I_5I_2$, and $\mathrm{res}(\varphi_1) = \mathrm{res}(\varphi_2) = \{I_4, [I_6I_1I_2]\}$. Note that $[I_6I_1I_3]$ and $[I_1I_3I_6]$ are the same circular string. Hence from the molecular point of view, both strategies yield the same molecules. Note also that in both strategies, two $\underline{\mathsf{ld}}$-rules were used, though they were applied to different set of pointers. The assembled sets are identical for the two strategies. $\qquad\square$

These examples lead to the following formulation of the notion of invariant.

A property $P$ is **invariant** (with respect to assembling) if for any two successful assembling strategies $\varphi_1$ and $\varphi_2$, if $\varphi_1(\delta)$ satisfies $P$, then so does $\varphi_2(\delta)$.

## 12.2 Invariant Theorem

The molecular $ld$ operation yields two molecules, a linear and a circular one. A gene (before excision and the addition of telomeres), as expressed by ($\ell$a1) and ($\ell$a2′), may end up assembled in a linear molecule or in a circular one. The formal results of this section indicate that whether or not a micronuclear gene is assembled in a linear or in a circular molecule depends *only* on the MDS/IES structure of the gene, and *not* on the choice of strategy to assemble the gene. Thus, if one strategy will place a gene in a circular molecule, then so do all other strategies. Also, the context of the assembled gene is invariant, i.e., the molecule containing the assembled gene will always contain exactly the same composite IES attached to it.

A crucial observation here is that during the assembly process, the markers are never separated from the IESs (and MDSs) adjacent to them. Also, the pointers are never separated from the IESs (and MDSs) adjacent to them, and moreover, all of them stop acting as pointers, after they become incorporated, at different moments of the assembly process, into bigger composite MDSs and IESs (each past pointer will have one occurrence in a composite IES and one in a composite MDS). Consequently, given a micronuclear gene, one can determine what composite IESs will be formed during any assembly process of that gene, thus giving in this way the above invariants.

We formalize these observations in the following two theorems.

**Theorem 12.2.** *Let $\delta$ be an MDS–IES descriptor. If $\varphi_1$ and $\varphi_2$ are any two successful assembly strategies for $\delta$, then $\varphi_1(\delta) = \varphi_2(\delta)$.*

Theorem 12.2 follows from the following theorem, which gives more detailed information concerning the assembled sets.

**Theorem 12.3.** *Let $\delta$ be an MDS–IES descriptor. If $\varphi_1$ and $\varphi_2$ are any two successful assembly strategies for $\delta$, then*

*(1) if $\varphi_1(\delta)$ is assembled in a linear string, then so is $\varphi_2(\delta)$;*
*(2) if $\varphi_1(\delta)$ is assembled in a linear string in orthodox order, then so is $\varphi_2(\delta)$;*
*(3) $\mathrm{con}(\varphi_1(\delta)) = \mathrm{con}(\varphi_2(\delta))$;*
*(4) $\mathrm{res}(\varphi_1(\delta)) = \mathrm{res}(\varphi_2(\delta))$;*
*(5) $\varphi_1$ and $\varphi_2$ have the same number of ld-rules.*

*Proof.* The following five claims are obvious from the definitions of ld, hi, and dlad on MDS–IES descriptors.

*Claim 1.* Applying simple $\mathrm{ld}_p$ to an MDS–IES descriptor $\delta$ implies that one IES sequence, left- and right-bordered by the two occurrences of $p$, is excised from $\delta$. All the other IES sequences remain unchanged.

*Claim 2.* Applying boundary $\mathrm{ld}_p$ to an MDS–IES descriptor $\delta$ has the following effects on the IES sequences of $\delta$:

(i) two IES sequences, one right-bordered by $p$, and the other left bordered by the other occurrence of $p$, are concatenated on those ends specified by $p$, preserving their relative order.
(ii) one IES sequence, left- and right-bordered by markers, is excised from the input string together with the adjoining MDSs, producing the assembled MDS–IES descriptor on a circular string: the ends bordered by the two occurrences of the markers are "joined together."

*Claim 3.* Applying $\mathrm{hi}_p$ to an MDS–IES descriptor $\delta$ implies that an IES sequence right-bordered in $\delta$ by $p$ or $\bar{p}$, is concatenated with an IES sequence left bordered by the same pointer, on the ends specified by that pointer. The two IES sequences preserve their relative order, and all the other IES sequences remain unchanged, modulo inversion.

*Claim 4.* Applying $\mathrm{dlad}_{p,q}$ to an MDS–IES descriptor $\delta$ has the following effects on the IES sequences of $\delta$:

(i) two IES sequences, one right-bordered by $p$ in $\delta$ and the other left-bordered by the other occurrence of $p$, get concatenated on the ends specified by $p$; moreover, they are not inverted and they preserve their relative order;
(ii) two IES sequences, one right-bordered by $q$ in $\delta$ and the other left-bordered by the other occurrence of $q$, get concatenated on the ends specified by $q$; moreover, they are not inverted and they preserve their relative order;
(iii) all the other IES sequences of $\delta$ remain unchanged.

*Claim 5.* If an IES sequence is left-bordered (right-bordered, resp.) bordered by a pointer $p$ in $\delta$, it will be left-bordered (right-bordered, resp.) by that pointer until the pointer is removed by one of the operations. If an IES sequence is bordered by a marker in $\delta$, then it will be bordered by that marker throughout all possible assembly strategies of $\delta$.

The following claim is a simple consequence of Claims 1–5.

*Claim 6.* Let $\delta$ be an MDS–IES micronuclear descriptor, and let $\varphi$ be a successful assembly strategy for $\delta$, and let $\iota$ be an IES sequence in $\delta$.

(i) If $\iota$ is left- and right-bordered in $\delta$ by a pointer $p$, then $[\iota]$ is in $\varphi(\delta)$.
(ii) If $\iota$ is left- and right-bordered in $\delta$ by markers, then $[(b,e)\iota]$ or $[(b,e)\bar{\iota}]$ is in $\varphi(\delta)$.
(iii) If $\iota$ is right-bordered by the marker $b$ ($\bar{e}$, resp.), but not left-bordered by the other marker, then $\iota(b,e)$ ($\iota(\bar{e},\bar{b})$, resp.) is a substring of a (possibly circular) string in $\varphi(\delta)$.
(iv) If $\iota$ is right bordered in $\delta$ by a pointer $p$, then there is a unique IES sequence $\rho$, left bordered in $\delta$ by $p$ such that either $\iota\rho$ or $\bar{\rho}\bar{\iota}$ is a substring of a string in $\varphi(\delta)$.

The following claim is now a simple consequence of Claim 6.

*Claim 7.* Let $\delta = (p_1, q_1)I_1(p_2, q_2)\ldots I_{n-1}(p_n, q_n)$ be an MDS–IES descriptor, and let $\varphi_1$ and $\varphi_2$ be any two successful assembly strategies for $\delta$. Then either the linear string is empty in both $\varphi_1(\delta)$ and $\varphi_2(\delta)$, or the first letter (which is an IES or $(b,e)$ or $(\bar{e},\bar{b})$) of the linear string is the same in $\varphi_1(\delta)$ and $\varphi_2(\delta)$.

*Proof of Claim 7.* Let $\varphi$ be any successful assembly strategy $\varphi$ for $\delta$. If $p_1$ is a pointer, then we distinguish two cases. If $q_n \neq p_1$, then clearly the IES that is in the first position of the linear string in $\varphi(\delta)$ is the IES that is left-bordered by $p_1$ in $\delta$. If $q_n = p_1$, then the linear string is empty.

On the other hand, if $p_1 = b$, then $(b, e)$ is in the first position of the linear string of $\varphi(\delta)$, and it is followed by the IES bordered by $e$ in $\delta$. A similar argument holds for the case where $p_1 = \bar{e}$, thus proving Claim 7.

By Claims 5, 6, and 7, it follows that the linear string of $\varphi(\delta)$ is the same for every successful assembly strategy $\varphi$ for $\delta$. This proves Case (1) of the theorem. Any IES not present in the linear string of $\varphi(\delta)$ must be excised in all assembly strategies for $\delta$. Thus, by Claim 6, this proves Cases (4) and (5) of the theorem. Moreover, it follows that the final place and orientation of the substring $(b, e)$ is also an invariant of the assembly, proving Case (2). Case (3) of the theorem then follows.                                                                        □

*Example 12.4.* Consider the actin gene in *Sterkiella nova*, having the MDS–IES descriptor

$$\delta = (3,4)I_1(4,5)I_2(6,7)I_3(5,6)I_4(7,8)I_5(9,e)I_6(\bar{3},\bar{2})I_7(b,2)I_8(8,9).$$

Consider then an assembly strategy for $\delta$, e.g., $\underline{\mathsf{ld}}_4$ $\underline{\mathsf{dlad}}_{5,6}$ $\underline{\mathsf{ld}}_7$ $\underline{\mathsf{dlad}}_{8,9}$ $\underline{\mathsf{hi}}_{\overline{2}}$ $\underline{\mathsf{hi}}_3$:

$$\underline{\mathsf{ld}}_4(\delta) = (3,5)I_2(6,7)I_3(5,6)I_4(7,8)I_5(9,e)I_6(\overline{3},\overline{2})I_7(b,2)I_8(8,9) \ + \ [I_1],$$

$$\underline{\mathsf{dlad}}_{5,6}(\underline{\mathsf{ld}}_4(\delta)) = I_0(3,7)I_3I_2I_4(7,8)I_5(9,e)I_6(\overline{3},\overline{2})I_7(b,2)I_8(8,9) \ + \ [I_1],$$

$$\underline{\mathsf{ld}}_7(\underline{\mathsf{dlad}}_{5,6}(\underline{\mathsf{ld}}_4(\delta))) = (3,8)I_5(9,e)I_6(\overline{3},\overline{2})I_7(b,2)I_8(8,9) \ + \ [I_1] \ + \ [I_3I_2I_4],$$

$$\underline{\mathsf{dlad}}_{8,9}(\underline{\mathsf{ld}}_7(\underline{\mathsf{dlad}}_{5,6}(\underline{\mathsf{ld}}_4(\delta)))) = (3,e)I_6(\overline{3},\overline{2})I_7(b,2)I_8I_5 \ + \ [I_1] \ + \ [I_3I_2I_4],$$

$$\underline{\mathsf{hi}}_{\overline{2}}(\underline{\mathsf{dlad}}_{8,9}(\underline{\mathsf{ld}}_7(\underline{\mathsf{dlad}}_{5,6}(\underline{\mathsf{ld}}_4(\delta))))) = (3,e)I_6(\overline{3},\overline{b})\overline{I}_7I_8I_5 \ + \ [I_1] \ + \ [I_3I_2I_4],$$

$$\underline{\mathsf{hi}}_3(\underline{\mathsf{hi}}_{\overline{2}}(\underline{\mathsf{dlad}}_{8,9}(\underline{\mathsf{ld}}_7(\underline{\mathsf{dlad}}_{5,6}(\underline{\mathsf{ld}}_4(\delta)))))) = \overline{I}_6(\overline{e},\overline{b})\overline{I}_7I_8I_5 \ + \ [I_1] \ + \ [I_3I_2I_4].$$

Thus, the gene is assembled in the inverted order, placed in a linear DNA molecule, with the IES $\overline{I}_6$ preceding it and the sequence of IESs $\overline{I}_7\,I_8\,I_5$ succeeding it. Two circular molecules are also produced: $[I_1]$ and $[I_3I_2I_4]$. □

*Example 12.5.* Consider the MDS–IES descriptor

$$\delta = (\overline{10},\overline{8})I_1(\overline{3},\overline{b})I_2(\overline{5},\overline{3})I_3(10,11)I_4(5,8)I_5(11,e)$$

of Example 12.2. Then $\delta$ is always assembled in a linear molecule, and no IES is excised during the assembly process, i.e., no $\underline{\mathsf{ld}}$ is ever applied in a process of assembling $\delta$. To prove this, it is enough to see that it holds for an arbitrary successful assembly strategy for $\delta$

$$\underline{\mathsf{hi}}_{\overline{10}}(\delta) = \overline{I}_3(3,5)\overline{I}_2(b,3)\overline{I}_1(8,11)I_4(5,8)I_5(11,e),$$

$$\underline{\mathsf{dlad}}_{8,11}(\underline{\mathsf{hi}}_{\overline{10}}(\delta)) = \overline{I}_3(3,5)\overline{I}_2(b,3)\overline{I}_1I_5I_4(5,e),$$

$$\underline{\mathsf{dlad}}_{3,5}(\underline{\mathsf{dlad}}_{8,11}(\underline{\mathsf{hi}}_{\overline{10}}(\delta))) = \overline{I}_3\overline{I}_1I_5I_4\overline{I}_2(b,e).$$

Moreover, the assembled descriptor will always be preceded by the IES sequence $\overline{I}_3\overline{I}_1I_5I_4\overline{I}_2$ and followed by the empty IES sequence. □

## Notes on References

- The results of this chapter are from Ehrenfeucht et al. [21].
- The proof of Theorem 12.3 does not rely essentially on our operations $\underline{\mathsf{ld}}$, $\underline{\mathsf{hi}}$, and $\underline{\mathsf{dlad}}$, but rather on the fact that every IES will always be concatenated with another IES through the same pointer. This must be so in any other model based on splicing, due to the very nature of the splicing operation. In the formalism of graphs this general invariant property is proved in Chap. 14.

# 13

# Patterns of Subsets of Rules

In Chap. 9 we showed that all legal strings are reducible to the empty string by a finite number of applications of the string rules snr, spr, and sdr corresponding to the molecular operations *ld*, *hi*, and *dlad*. In particular, this is true for all realistic legal strings. A central question for understanding the role of this set of operations in the process of assembly is the following. What kind of realistic MDS arrangements (corresponding to realistic legal strings) can be successfully assembled by various subsets of this set?

In this chapter we provide a complete answer to this question: for each subset $S$ of the rules snr, spr, sdr, we characterize the realistic legal strings that are successful in the string pointer reduction system using only operations from the subset $S$. In particular, this will naturally enable us to introduce both complexity and similarity measures on the strings. For instance, a realistic MDS arrangement that can be assembled by using only the operations snr and spr can be regarded as "easier" than a pattern that requires the more complex operation sdr.

## 13.1 Small Reductions

In this chapter, we shall consider legal strings $v$ of pointers, $v \in \Delta_\kappa^\circledast$ with $\kappa \geq 2$. Let $S$ be a nonempty set of operations in $\mathsf{Snr} \cup \mathsf{Spr} \cup \mathsf{Sdr}$. A composition of operations from $S$ is an $S$-**reduction**. A legal string $v \in \Delta_\kappa^\circledast$ is said to be **successful in** $S$ if there exists an $S$-reduction $\varphi$ such that $\varphi(v) = \Lambda$. By Theorem 9.1, every legal string is successful in the full set $S = \mathsf{Snr} \cup \mathsf{Spr} \cup \mathsf{Sdr}$ of the three operations. The problem that thus remains to be considered is the case of proper nonempty subsets of operations. We shall first consider the set of legal strings having legal substrings, reducing the characterization problem of subsets to two smaller instances of the same problem.

**Lemma 13.1.** *Let $u = w_1 v w_2$ be a legal string, where the substring $v$ is legal. If $\varphi_1, \varphi_2 \in \mathsf{Snr} \cup \mathsf{Spr} \cup \mathsf{Sdr}$ such that $\varphi_1$ is applicable to $v$ and $\varphi_2$ is applicable to $w_1 w_2$, then $\varphi_1 \varphi_2(u) = \varphi_2 \varphi_1(u)$.*

*Proof.* Notice first that $w_1w_2$ is also a legal string. Since $\varphi_1$ is applicable to $v$, the strings $w_1$ and $w_2$ are not changed in $\varphi_1(w_1vw_2)$; that is, $\varphi_1(u) = w_1\varphi_1(v)w_2$. Let $w_1'$ and $w_2'$ result from the substrings $w_1$ and $w_2$ when $\varphi_2$ is applied to $u$, i.e., $\varphi_2(w_1w_2) = w_1'w_2'$. Now,

$$\varphi_2\varphi_1(u) = \varphi_2(w_1\varphi_1(v)w_2) = w_1'\varphi_1(v)^e w_2',$$

where $\varphi_1(v)^e$ is either $\varphi_1(v)$ or $\overline{\varphi_1(v)}$ depending on $\varphi_2$ and $w_1w_2$, since the operations in $\varphi_2$ can invert the string $\varphi_1(v)$, but cannot change it otherwise. Also,

$$\varphi_1\varphi_2(u) = \varphi_1(w_1'v^e w_2') = w_1'\varphi_1(v^e)w_2' = w_1'\varphi_1(v)^e w_2',$$

and thus $\varphi_1\varphi_2(u) = \varphi_2\varphi_1(u)$, as required. □

Let $u = w_1vw_2$ be a legal string such that the substring $v$ is legal. If $\varphi(u) = \Lambda$ for a string reduction $\varphi$, then, by Lemma 13.1, every string rule $\varphi_1 \in \mathsf{Snr} \cup \mathsf{Spr} \cup \mathsf{Sdr}$ in $\varphi = \varphi''\varphi_1\varphi_2\varphi'$ that is applicable to $v$ can be commuted with the neighboring rule $\varphi_2$ that is applicable to $w_1w_2$. This observation yields the following lemma.

**Lemma 13.2.** *Let $\mathcal{S}$ be a subset of operations, and let $u = w_1vw_2$ be a legal string, where $v$ is also legal. Suppose that an $\mathcal{S}$-reduction $\varphi$ is applicable to $u$. Then there are $\mathcal{S}$-reductions $\varphi_1$ and $\varphi_2$ such that $\varphi(u) = \varphi_2\varphi_1(u)$, and $\varphi_1$ is applicable to $v$, and $\varphi_2$ is applicable to $w_1w_2$.*

*Example 13.1.* Consider the legal string $u = 2\,3\,5\,\overline{4}\,4\,5\,\overline{2}\,3$. Then $u = w_1vw_2$, where both strings $v = 5\,\overline{4}\,4\,5$ and $w_1w_2 = 2\,3\,\overline{2}\,3$ are legal. We have

$$u_1 = \mathsf{spr}_3(u) = 2\,2\,\overline{5}\,\overline{4}\,4\,5, \qquad u_2 = \mathsf{spr}_{\overline{4}}(u_1) = 2\,2\,\overline{5}\,5,$$
$$u_3 = \mathsf{snr}_2(u_2) = \overline{5}\,5, \qquad \mathsf{snr}_{\overline{5}}(u_3) = \Lambda.$$

Therefore, $\varphi = \mathsf{snr}_{\overline{5}}\,\mathsf{snr}_2\,\mathsf{spr}_{\overline{4}}\,\mathsf{spr}_3$ is successful for $u$. The operations that apply to $v$ form the composition $\varphi_1 = \mathsf{snr}_{\overline{5}}\,\mathsf{spr}_{\overline{4}}$, and those that apply to $w_1w_2$ form the composition $\varphi_2 = \mathsf{snr}_2\,\mathsf{spr}_3$. We conclude that $\varphi_1(v) = \mathsf{snr}_{\overline{5}}\,\mathsf{spr}_{\overline{4}}(v) = \Lambda$ and $\varphi_2(w_1w_2) = \mathsf{snr}_2\,\mathsf{spr}_3(w_1w_2) = \Lambda$, and therefore, $\varphi_2\varphi_1 = \mathsf{snr}_2\,\mathsf{spr}_3\,\mathsf{snr}_{\overline{5}}\,\mathsf{spr}_{\overline{4}}$ is successful for $u$. □

Using Lemma 13.2 we have

**Lemma 13.3.** *Let $\mathcal{S}$ be a subset of operations, and let $u = w_1vw_2$ be a legal string, where $v$ is also legal. Then $u$ is successful in $\mathcal{S}$ if and only if both $v$ and $w_1w_2$ are successful in $\mathcal{S}$.*

*Proof.* Suppose first that $v$ and $w_1w_2$ are successful in $\mathcal{S}$. Then there are $\mathcal{S}$-reductions $\varphi_1$ and $\varphi_2$ such that $\varphi_1(v) = \Lambda$ and $\varphi_2(w_1w_2) = \Lambda$. It is then clear that $\varphi_2\varphi_1(v) = \Lambda$, and therefore $v$ is successful in $\mathcal{S}$.

Conversely, suppose $u$ is successful in $\mathcal{S}$, and let $\varphi$ be an $\mathcal{S}$-reduction such that $\varphi(u) = \Lambda$. By Lemma 13.2, $\varphi(u) = \varphi_2\varphi_1(u)$, where $\varphi_1$ and $\varphi_2$ are $\mathcal{S}$-reductions such that $\varphi_1$ is applicable to $v$ and $\varphi_2$ is applicable to $w_1w_2$. It is then clear that $\varphi_1(v) = \Lambda$, since $\varphi_2\varphi_1(u) = \Lambda$ and none of the operations taking part in the composition $\varphi_2$ are applicable to $v$. Since $\varphi_1(u) = w_1 w_2$, it follows that also $\varphi_2(w_1 w_2) = \Lambda$. $\qquad\square$

We note also the following property related to the successfulness of a realizable word and that of any of its conjugates:

**Lemma 13.4.** *Let $u$ be a realizable legal word and $\mathcal{S}$ a subset of operations. Then $u$ is successful in $\mathcal{S}$ if and only if any conjugate of $u$ is successful in $\mathcal{S}$.*

*Proof.* Clearly, any two conjugated legal strings have the same overlap graph. The lemma then follows from Theorems 11.2 and 11.3. $\qquad\square$

## 13.2 Disjoint Cycles

We shall now look how the IESs are joined together in a successful assembly strategy. In a molecular assembly strategy the MDSs and IESs can form circular molecules when the operation $ld$ is applied to a linear molecule. As we saw in Chap. 12, these circular molecules are independent of the chosen assembly strategy. In this section we shall model the formation of circular molecules by *IES cycles* in MDS descriptors and by *disjoint cycles* in strings.

Let $\alpha = N_1 N_2 \ldots N_k$ be a realistic (MDS) arrangement of size $\kappa$, where $\psi_\kappa(N_i) = (p_i, q_i) \in \Gamma_\kappa \cup \overline{\Gamma}_\kappa$ for each $i$. We denote by $I_i = (q_i, p_{i+1})$ the pair that represents the IES interlacing the MDSs $N_i$ and $N_{i+1}$ in $\alpha$ for $i \in [1, k]$. Also, we say that $I_k = (q_n, p_1)$ is the **connector** of $\alpha$. It is the pair formed by the right pointer (or marker) of $N_k$ and the left pointer (or marker) of $N_1$. We call the sequence $\iota(\alpha) = I_1 I_2 \ldots I_k$ of pairs the **IES descriptor** of $\alpha$. It is clear that $\alpha$ is uniquely determined by its IES descriptor $\iota(\alpha)$.

*Example 13.2.* (1) The arrangement $\alpha = M_{2,4}\overline{M}_{7,9}M_{1,1}M_{5,6}$ is realistic of size $\kappa = 9$. Its MDS descriptor is $\delta = (2,5)(\overline{e},\overline{7})(b,2)(5,7)$, and its IES descriptor is $\iota(\alpha) = (5,\overline{e})(\overline{7},b)(2,5)(7,2)$. Together with the size $\kappa$, the IES descriptor determines $\alpha$.

(2) Let $\alpha = M_1\overline{M}_3 M_4 M_2$ be a micronuclear arrangement of size $\kappa = 4$. Its MDS descriptor is $\delta = (b,2)(\overline{4},\overline{3})(4,e)(2,3)$, and its IES descriptor is $\iota(\alpha) = (2,\overline{4})(\overline{3},4)(e,2)(3,b)$.

(3) Let $\alpha = \overline{M}_3 M_2 \overline{M}_1$. Then its MDS descriptor is $\delta = (\overline{e},\overline{3})(2,3)(\overline{2},\overline{b})$, and its IES descriptor is $\iota(\alpha) = (\overline{3},2)(3,\overline{2})(\overline{b},\overline{e})$. $\qquad\square$

A pointer $p$ **occurs** in an IES descriptor $\iota$ if there is a pair $\iota'$ in $\iota$ where either one component is equal to $p$ or its inversion $\overline{p}$.

Let $\alpha$ be a realistic arrangement, and let $\iota$ be its IES descriptor. An **IES cycle of** $\iota$ (as well as of $\alpha$) is any proper scattered substring $\iota'$ of $\iota$ such that

- if a pointer $p$ (or $\bar{p}$) occurs in $\iota'$, then there are two occurrences from **p** in $\iota'$, and
- if a marker $b$ or $e$ (or an inversion of a marker) occurs in $\iota'$, then both markers occur in $\iota'$.

In this case we also say that $\iota$ **contains** $\iota'$, and that $\iota$ and $\alpha$ are **circular**. If $\iota$ does not contain $\iota'$, then we say that $\iota$ **avoids** $\iota'$.

An IES descriptor $\iota$ can be thought of as a graph, where the vertices are the pointers occurring in $\iota$, and each pair $(p, q)$ in $\iota$ gives an (undirected) edge between the pointers $p$ and $q$. Then an IES cycle corresponds either to a cycle in this graph or to a path between the markers $b$ and $e$. An IES cycle is, however, supposed not to consist of the whole graph. In particular, if $\iota'$ is an IES cycle, then $\mathrm{rem}_\kappa(\iota')$ is a legal string, where the mapping $\mathrm{rem}_\kappa$ removes the parentheses and the markers from the strings of pairs (see p. 84).

The IES structure $\iota$ of $\alpha$ (as well as $\alpha$, itself) is said to be a **linear structure** if $\iota$ contains no IES cycles.

*Example 13.3.* (1) Consider the arrangement $\alpha = M_1 M_4 \overline{M}_2 \overline{M}_5 M_7 M_3 M_6$. Then the IES descriptor of $\alpha$ is $\iota = (2, 4)(5, \overline{3})(\overline{2}, \overline{6})(\overline{5}, 7)(e, 3)(4, 6)(7, b)$. It is circular, since it contains the scattered substring $\iota' = (2, 4)(\overline{2}, \overline{6})(4, 6)$, which is an IES cycle (of $\iota$). Observe that the rest of the IES descriptor, i.e., $(5, \overline{3})(\overline{5}, 7)(e, 3)(7, b)$, is also necessarily circular. The graph illustrating these facts is given in Fig. 13.1.

(2) The micronuclear arrangement $\alpha$ in Example 13.2(2) is linear. Indeed, assume that the IES descriptor $\iota$ of $\alpha$ contains an IES cycle $\iota'$. Then 2 occurs in $\iota'$ if and only if 4 and $e$ also occur in $\iota'$, and this holds if and only if 3 occurs in $\iota'$. But then $\iota' = \iota$, and hence $\iota'$ is not an IES cycle of $\iota$.    □

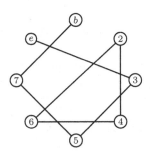

**Fig. 13.1.** The graph on the vertices $\{b, 2, \ldots, 7, e\}$ illustrating the IES descriptor $\iota = (2, 4)(5, \overline{3})(\overline{2}, \overline{6})(\overline{5}, 7)(e, 3)(4, 6)(7, b)$ from Example 13.3. An IES cycle corresponds either to a cycle in this graph or to a path between the markers. Hence both $\iota' = (2, 4)(\overline{2}, \overline{6})(4, 6)$ and its "complement" $\iota'' = (5, \overline{3})(\overline{5}, 7)(e, 3)(7, b)$ are IES cycles in $\iota$.

The property of a micronuclear arrangement $\alpha$ to contain an IES cycle $\iota'$ turns out to be instrumental in the following where we characterize the realistic legal strings corresponding to micronuclear arrangements that can be assembled without the use of string negative rules $snr_p$. The required property of micronuclear arrangements can be translated to the level of realistic legal strings using the notions of *legal* sets of strings and *disjoint cycles*, which are defined in the following.

Let $P = \{v_1, v_2, \ldots, v_m\}$, for $m \geq 1$, be a set of signed strings over $\Delta_\kappa$ for some $\kappa \geq 2$. We say that $P$ is **legal** if the string $w_P = v_1 v_2 \ldots v_m$ is legal. Notice that the notion of a legal set is well defined, since if $w_P$ is legal, so are the strings $v_{i_1} \ldots v_{i_m}$ for all permutations $(i_1 i_2 \ldots i_m)$ of $[1, m]$. A pointer $p$ is said to **occur** in $P$ if it occurs in $w_P$. Moreover, $p$ is **negative in $P$** (**positive in $P$**, resp.), if it is negative (positive, resp.) in $w_P$. Obviously, also the qualification of being positive or negative is independent of the order of $P$.

A legal set $P$ is a **disjoint cycle** if it can be ordered, $v_1, v_2, \ldots, v_m$, so that the following conditions are satisfied for some $p_1, \ldots, p_m \in \Pi_\kappa$:

$$\left\{ \begin{array}{l} v_1 = p_1 p_2, \\ v_{i-1} = p_{i-1} p_i \implies v_i = p_i p_{i+1}, \quad \text{or} \quad v_i = p_{i+1} \bar{p}_i, \\ v_{i-1} = p_i p_{i-1} \implies v_i = p_{i+1} p_i, \quad \text{or} \quad v_i = \bar{p}_i p_{i+1}, \end{array} \right. \tag{13.1}$$

where $p_{m+1} = p_1$. Note that in a disjoint cycle $P$ all element strings have length two.

*Example 13.4.* The sets $P_1 = \{23, 35, 54, 42\}$ and $P_2 = \{2\bar{4}, \bar{5}4, 56, 62\}$ are disjoint cycles. The set $P_3 = \{23, 34, \bar{4}\bar{2}\}$ is not, since after $v_{i-1} = 34$ (i.e., $p_i = 4$) should follow a string $v_i = 4p_{i+1}$ or $v_i = p_{i+1}\bar{4}$, but instead $\bar{4}\bar{2}$ can be found in $P_3$. □

Note that if $P$ is a disjoint cycle with respect to an ordering $v_1, \ldots, v_m$ as required in (13.1), then all the cyclic permutations $v_i, \ldots, v_m, v_1, \ldots, v_{i-1}$ of this order will also satisfy conditions (13.1).

Notice also that in the special case for $m = 1$, the disjoint cycles are the legal sets $P = \{pp\}$ for pointers $p$. The following lemma follows directly from the definition.

**Lemma 13.5.** *If all pointers of a legal set $P$ are negative in $P$, then $P$ is a disjoint cycle if and only if $P = \{p_1 p_2, p_2 p_3, \ldots, p_m p_1\}$ for some $p_1, p_2, \ldots, p_m \in \Pi_\kappa$ with $m \geq 1$.*

Let $P = \{v_1, v_2, \ldots, v_m\}$ be a disjoint cycle and $u$ a legal string. We say that $u$ **contains** $P$ if there exists a conjugate $u'$ of $u$ such that $u' = u_1 v_{i_1} u_2 v_{i_2} \ldots u_m v_{i_m} u_{m+1}$ for some substrings $u_j$, for $j \in [1, m+1]$, and a permutation $(i_1 i_2 \ldots i_m)$ of $[1, m]$. Thus the elements $v_1, \ldots, v_m$ occur in $u'$ without overlapping each other. If $u$ does not contain $P$, then we say that $u$ **avoids** $P$.

*Example 13.5.* (1) Consider disjoint cycles $P_1 = \{23, 35, 54, 42\}$ and $P_2 = \{2\overline{4}, \overline{5}4, 56, 62\}$ from Example 13.4. The string $u = 5\,2\,3\,6\,4\,2\,5\,4\,6\,3$ contains $P_1$, since the elements of $P_1$ occur in the conjugate $u' = 2\,3\,6\,4\,2\,5\,4\,6\,3\,5$ without overlapping each other: $u' = \mathbf{23}\,6\,\mathbf{42}\,\mathbf{54}\,6\,\mathbf{35}$. On the other hand, $u$ avoids $P_2$.

(2) Consider the legal string $u = 2\,3\,5\,6\,2\,7\,4\,5\,\overline{4}\,\overline{3}\,7\,\overline{6}$. Then $u$ contains the disjoint cycle $P = \{35, \overline{5}4, 74, \overline{3}7\}$: $u = 2\,\mathbf{35}\,6\,2\,\mathbf{74}\,\mathbf{54}\,\overline{\mathbf{37}}\,\overline{6}$. This cycle is illustrated graphically as a cycle in Fig. 13.2. The vertices of the graph are the occurrences of the pointers in $u$, and the edges are the pairs in $P$ as well as the vertices that have as their label the same pointer $p$ or $\overline{p}$.  □

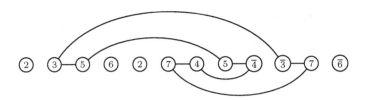

**Fig. 13.2.** An illustration of the disjoint cycle in Example 13.5(2) for the legal string $u = 2\,3\,5\,6\,2\,7\,4\,5\,\overline{4}\,\overline{3}\,7\,\overline{6}$. Here $u$ contains the disjoint cycle $P = \{35, \overline{5}4, 74, \overline{3}7\}$, which can traced as a cycle in the graph by following the edges.

We are now ready to prove the connection between disjoint cycles of strings and IES cycles of MDS arrangements.

**Lemma 13.6.** *Let $\alpha$ be a realistic arrangement and let $u = \varrho_\kappa(\alpha)$ be the corresponding realistic legal string. Then $u$ contains a disjoint cycle if and only if $\alpha$ contains an IES cycle.*

*Proof.* By Lemma 6.1, for each realistic arrangement $\alpha$, there corresponds a micronuclear arrangement $\alpha'$ such that $u = \varrho_\kappa(\alpha)$ and $u' = \varrho_\kappa(\alpha')$ are isomorphic. It is clear that $\alpha$ contains an IES cycle if and only if $\alpha'$ does, and that the string $u$ contains a disjoint cycle if and only if $u'$ contains such a cycle. Therefore, we can assume that $\alpha$ is a micronuclear arrangement and, consequently, $\mathrm{dom}(u) = [2, \kappa]$ for some $\kappa \geq 2$.

Let $\iota = \iota(\alpha)$ be the IES descriptor of $\alpha$. Assume first that $\iota$ (and thus $\alpha$) has an IES cycle $\iota'$. If the markers do not occur in $\iota'$, then the set $P = \{pq \mid (p, q)$ is in $\iota'\}$ is a disjoint cycle of $u$. Indeed, suppose that $(p, q)$ is in $\iota$. Then

(1) either $(q, t)$ or $(t, \overline{q})$ is in $\iota$ for some $t$, and
(2) either $(s, p)$ or $(\overline{p}, s)$ is in $\iota$ for some $s$.

From these cases the conditions (13.1) can be easily verified for $P$. On the other hand, if the markers do occur in $\iota'$, then the IESs that do not occur in $\iota'$ form another IES cycle of $\iota$, and so the set

$$P' = \{pq \mid (p,q) \text{ is in } \iota, \text{ but not in } \iota'\}$$

is a disjoint cycle of $u$.

Assume then that $u$ contains a disjoint cycle $P$. If there is a string $pq \in P$ and an MDS $M$ of $\alpha$ such that $\varrho(M) = pq$, then necessarily

$$P = \{r_1 r_2 \mid (r_1, r_2) = (i, i+1) \text{ or } (r_1, r_2) = (\overline{i+1}, \overline{i}) \text{ for some } 2 \le i < \kappa\}$$
$$\cup \ \{r_1 r_2 \mid (r_1, r_2) = (\kappa, 2) \text{ or } (r_1, r_2) = (\overline{2}, \overline{\kappa})\}.$$

It follows from the second set of this formula that either $M_\kappa M_1$ or $\overline{M}_1 \overline{M}_\kappa$ is a subsequence of $\alpha$, and therefore either $(e, b)$, or $(\overline{b}, \overline{e})$ is an IES of $\iota$. Since both of them are circular, $\alpha$ has an IES cycle.

Finally, suppose that there exists no string $pq \in P$ such that $\varrho(M) = pq$, for any MDS $M$ of $\alpha$. Then necessarily, the IESs in the set $\{(p, q) \mid pq \in P\}$ form an IES cycle of $\alpha$. $\qquad\square$

We show now how disjoint cycles are related to the operation snr: in order to reduce a realistic legal string containing a disjoint cycle to the empty string, an operation from Snr must be used at least once.

**Lemma 13.7.** *Let $u$ be a legal string that contains a disjoint cycle $P$.*

*(a) If $\mathsf{spr}_p$ is applicable to $u$, then $\mathsf{spr}_p(u)$ contains a disjoint cycle.*
*(b) If $\mathsf{sdr}_{p,q}$ is applicable to $u$, then $\mathsf{sdr}_{p,q}(u)$ contains a disjoint cycle.*

*Proof.* Assume that $\mathsf{spr}_p$ is applicable to $u$, say $u = u_1 p u_2 \overline{p} u_3$. We can assume that $p$ occurs in $P$; otherwise, $P$ is a disjoint cycle in $\mathsf{spr}_p(u)$. Therefore, there are pointers $r$ and $s$ such that either $A = \{rp, s\overline{p}\} \subseteq P$, or $A = \{pr, \overline{p}s\} \subseteq P$. In both cases, it is clear that $(P \setminus A) \cup \{rs\}$ is a disjoint cycle in $\mathsf{spr}_p(u)$. This proves Case (a).

Let then $p$ and $q$ be pointers in $u$ such that $u = u_1 p u_2 q u_3 p u_4 q u_5$ for some substrings $u_i$, and let $v = \mathsf{sdr}_{p,q}(u)$. If $p$ and $q$ do not occur in $P$, then obviously $v$ contains the disjoint cycle $P$. If only one of them, say $p$, occurs in $P$, then there are pointers $r, s \notin \{p, q\}$ such that $rp, ps \in P$. Now also $Q = (P \setminus \{rp, ps\}) \cup \{rs\}$ is a disjoint cycle, and $v = u_1 u_4 u_3 u_2 u_5$ contains $Q$.

If both $p$ and $q$ occur in $P$, then we have three possibilities: either (i) $pq, qp \notin P$, or (ii) $pq \in P$ and $qp \notin P$, or (iii) $qp \in P$ and $pq \notin P$.

If (i) holds, there are pointers $r_1, r_2, s_1, s_2$ such that $r_1 p, pr_2, s_1 q, qs_2 \in P$. Thus $v$ contains the disjoint cycle $(P \setminus \{r_1 p, pr_2, s_1 q, qs_2\}) \cup \{r_1 r_2, s_1 s_2\}$.

If (ii) holds, there are pointers $r, s$ such that $rp, pq, qs \in P$. Thus $v$ contains the disjoint cycle $(P \setminus \{rp, pq, qs\}) \cup \{rs\}$.

If (iii) holds, there are pointers $r, s$ such that $rq, qp, ps \in P$. Thus $v$ contains the disjoint cycle $(P \setminus \{rq, qp, ps\}) \cup \{rs\}$. $\qquad\square$

Lemma 13.7 yields the following corollary.

**Corollary 13.1.** *Let $u$ be a legal string containing a disjoint cycle, and let $\varphi$ be a successful string reduction for $u$. Then $\varphi$ contains a string negative rule $\mathsf{snr}_p$ for some pointer $p$.*

## 13.3 Subsets of Successful Patterns

In the following subsections we shall systematically consider the cases of all different unions of the sets Snr, Spr, and Sdr for $S$. For each of these subsets $S$, we give a characterization of the realistic legal strings that are successful in $S$.

### 13.3.1 snr

As the first case we consider the realistic legal strings that are successful using only the string negative rule: $S = $ Snr. We show that these strings are characterized by the realistic conjugates of the string $2233\ldots\kappa\kappa$.

**Theorem 13.1.** *A realistic legal string $u$ is successful in* Snr *if and only if $u$ or $\overline{u}$ is isomorphic to a string in*

$$L = \{2233\ldots\kappa\kappa\} \cup \{p(p+1)(p+1)\ldots\kappa\kappa\,22\ldots(p-1)(p-1)\,p \mid 2 \le p \le \kappa\}.$$

*Proof.* Since the claim allows isomorphism, we can assume that $\mathrm{dom}(u) = [2, \kappa]$, i.e., that $u = \varrho_\kappa(\alpha)$ for a micronuclear arrangement $\alpha$ (instead of for a general realistic arrangement $\alpha$).

First, we notice that if $u \in L$ or $\overline{u} \in L$, then clearly $\varphi(u) = \Lambda$ for the composition

$$\varphi = \mathsf{snr}_p\,\mathsf{snr}_{p+1}\ldots\mathsf{snr}_\kappa\,\mathsf{snr}_{p-1}\ldots\mathsf{snr}_2\,.$$

Conversely, assume that $u$ is a realistic legal string successful in Snr. Then all pointers of $u$ are negative, i.e., either $u \in \Delta_\kappa^*$ or $u \in \overline{\Delta}_\kappa^*$. Clearly, if $u$ is successful in Snr, then so is $\overline{u}$, and therefore, without loss of generality, we can assume that $u \in \Delta_\kappa^*$. Recall that, for each pointer $p$, $u = u'u_{(p)}u''$, where $u_{(p)}$ is the interval bordered by the two occurrences of $p$. In the string $u$ no two pointers overlap, since an application of a string negative rule can never resolve overlapping of pointers. It follows that if, for some pointer $q$, one occurrence of $q$ is in $u_{(p)}$, then so is the other. Hence $u_{(p)}$ is a legal substring of $u$ for each $p$.

If $u_{(p)} = pp$ for all pointers $p$, then obviously $u = 2233\ldots\kappa\kappa$. Suppose then that there is a pointer $p$ such that $u_{(p)} \ne pp$, and let $q \ne p$ be in $u_{(p)}$. Let us assume that $q < p$; the other case is solved similarly. Since $u$ is realistic, we have from Lemma 10.3(i) that $2 \in \mathrm{dom}(u_{(p)})$, and hence also that $[2, p] \subseteq \mathrm{dom}(u_{(p)})$. In particular, $p - 1$ occurs in the substring $(p-1)p$ at the end of $u_{(p)}$, and then $p + 1$ (if $p < \kappa$) must occur in the substring $p(p+1)$ in the beginning of $u_{(p)}$. Now, Lemma 10.3 implies that $\mathrm{dom}(u_{(p)}) = [2, \kappa]$, i.e., $u = u_{(p)}$. Clearly, now $p$ is unique with the property that $u_{(p)} \ne pp$, and hence $u_{(p)}$ is a concatenation of the strings $qq$ for all $q$ with $q \ne p$. As $u$ is realistic, we conclude that $u = p(p+1)(p+1)(p+2)\ldots(\kappa-1)\kappa\kappa\,223\ldots(p-1)p$ as required.    $\square$

*Example 13.6.* The R1 gene of *Sterkiella nova* is described by the micronuclear arrangement $M_1 M_2 M_3 M_4 M_5 M_6$. The corresponding realistic legal string 2233445566 can be successfully reduced using the string negative rules, while the realistic legal string 2323 $(= \varrho_3(M_1 M_3 M_2))$ cannot be assembled in this way.                                                                                                        □

Note that a realistic legal string $u$ is reducible in Snr if and only if the overlap graph $\gamma_u$ has only isolated vertices all of which are negative. Thus the corresponding result for overlap graphs is very simple.

### 13.3.2 snr and spr

We need the following notation for the case where we allow string negative and string positive rules but no string double rules. A legal string is **elementary** if it has no proper legal substrings.

*Example 13.7.* The string $v = 2\,\overline{5}\,4\,3\,4\,5\,3\,\overline{2} \in \Delta_5^{\maltese}$ is legal, but it is not elementary, since it has a proper substring $\overline{5}\,4\,3\,4\,5\,3$ that is also legal.    □

**Lemma 13.8.** *Let $u$ be an elementary legal string.*

*(i) If $u$ is of length two, then $u$ is successful in $\mathsf{Snr} \cup \mathsf{Spr}$.*

*(ii) If $u$ is longer than two, then $u$ is successful in $\mathsf{Snr} \cup \mathsf{Spr}$ if and only if $u$ contains at least one positive pointer.*

*Proof.* Case (i) is obvious, since the only legal strings of length two are isomorphic to 22, $2\overline{2}$, $\overline{2}2$, and $\overline{2}\overline{2}$. Assume then that the $u$ is longer than two.

If $u$ is successful in $\mathsf{Snr} \cup \mathsf{Spr}$, then clearly $u$ contains at least one positive pointer, since otherwise $u$ is successful in Snr, but, by Theorem 13.1, $u$ would then have no proper legal substrings. Thus the condition in the lemma is necessary.

Let us assume that the condition is not sufficient, and let $u$ be a minimal elementary legal string having positive pointers, but unsuccessful in $\mathsf{Snr} \cup \mathsf{Spr}$. Let $S$ be the set of the positive pointers of $u$ that have a maximum number of overlapping negative pointers in $u$. Also, let $S' \subseteq S$ consist of the pointers in $S$ that have the minimum number of positive overlapping pointers in $u$. Let $p \in S'$ be the leftmost pointer in $u$. In particular, $\mathsf{spr}_p$ is applicable to $u$, since $p$ is positive. Since $u$ is assumed to be a minimal counterexample, Lemma 13.3 implies that $\mathsf{spr}_p(u)$ has a legal substring $v$ consisting of negative pointers.

We shall prove now that $v$ is reducible in Snr. Let $A$ be the set of pointers in $v$. Since $u$ is elementary, it has no legal substrings, and thus $v$ must contain a positive pointer $q$ that overlaps with $p$, that is, $q \in O_u^+(p)$. Now, the negative pointers of $u$ that overlap with $p$ must also overlap with $q$, since $v$ does not contain positive pointers. Therefore $O_u^-(p) \subseteq O_u^-(q)$. Since $\mathrm{card}(O_u^-(p))$ is maximal, we have $O_u^-(p) = O_u^-(q)$. By the same argument, also $O_u^+(q) \subseteq O_u^+(p)$, and hence $O_u^+(p) = O_u^+(q)$. Thus $O_u(p) = O_u(q)$ for all $q \in O_u^+(p)$ such that $q$ occurs in $v$ in $\mathsf{spr}_p(u)$.

Clearly, $A \cap O_u(p) \subseteq O_u^+(p)$. By the above, for any $q \in A \cap O_u(p)$, $O_u(q) = O_u(p)$. On the other hand, since $u$ has no legal substrings, any two pointers of $u$ are connected in the overlap graph $\gamma_v$ (that is, for any two pointers $r$ and $t$, there is a sequence $r_1, \ldots, r_n$ with $r_1 = r$, $r_n = t$, and $r_{i+1} \in O_u(r_i)$ for all $i$). Thus there are no pointers $r \in A \setminus O_u(p)$, and consequently, $A \subseteq O_u^+(p)$ in $u$. Let $A = \{q_1, \ldots, q_i\}$ for some $i \geq 1$. Since $O_u(p) = O_u(q_k)$ for all $1 \leq k \leq i$ (and $u$ is elementary), it follows that, for some strings $u_1, u_2, u_3$,

$$u = u_1 p\, q_1 \ldots q_i u_2 \overline{p}\, \overline{q}_1 \ldots \overline{q}_i u_3.$$

Thus, any negative legal substring of $\mathsf{spr}_p(u)$ is of the form $\overline{q}_i \ldots \overline{q}_1 \overline{q}_1 \ldots \overline{q}_i$, for some $i \geq 1$. Consequently, any negative legal substring of $\mathsf{spr}_p(u)$ is successful in $\mathsf{Snr}$ and thus, also in $\mathsf{Snr} \cup \mathsf{Spr}$. However, $\mathsf{spr}_p(u)$ may have nonnegative legal substrings. As $u$ was a minimal counterexample, all such substrings are successful in $\mathsf{Snr} \cup \mathsf{Spr}$. This is a contradiction, as it implies that $u$ is successful in $\mathsf{Snr} \cup \mathsf{Spr}$. $\qquad \square$

The following result follows then by Lemma 13.3 and Lemma 13.8.

**Theorem 13.2.** *A realistic string $u$ is successful in $\mathsf{Snr} \cup \mathsf{Spr}$ if and only if for all legal substrings $v$ of $u$, if $v = v_1 u_1 v_2 \ldots v_j u_j v_{j+1}$, where each $u_i$ is a legal substring, then $v_1 v_2 \ldots v_{j+1}$ contains a positive pointer or it is reducible in $\mathsf{Snr}$.*

The situation in Theorem 13.2 is illustrated in Fig. 13.3.

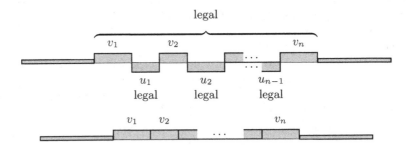

**Fig. 13.3.** Illustration of Theorem 13.2. A realistic string $u$ is not successful in $\mathsf{Snr} \cup \mathsf{Spr}$ if it has a legal substring $v = v_1 u_1 v_2 \ldots v_j u_j v_{j+1}$ for legal $u_i$ such that $v_1 v_2 \ldots v_{j+1}$ does not contain a positive pointer or it is not successful in $\mathsf{Snr}$.

*Example 13.8.* (1) The realistic string $2\,\overline{3}\,\overline{2}\,3$ corresponding to the micronuclear arrangement $M_1 \overline{M}_2 M_3$ is successful in $\mathsf{Snr} \cup \mathsf{Spr}$, but it is not successful in either $\mathsf{Snr}$ or $\mathsf{Spr}$.

(2) The realistic legal string $u = 3\,4\,2\,7\,8\,4\,5\,6\,\overline{5}\,\overline{6}\,6\,7\,9\,8\,9\,2\,3$, corresponding to the MDS arrangement $M_3 M_1 M_7 M_4 \overline{M}_5 M_6 M_9 M_8 M_2$, is not successful in $\mathsf{Snr} \cup \mathsf{Spr}$. This is because if we choose $v = u$ in Theorem 13.2 and consider the decomposition $v = v_1 u_1 v_2$ for $v_1 = 3\,4\,2\,7\,8\,4$, $u_1 = 5\,6\,\overline{5}\,6$ and $v_2 = 7\,9\,8\,9\,2\,3$, then $v_1 v_2$ is legal, it has no positive pointers and it is not reducible in $\mathsf{Snr}$.    □

Note that a realistic legal string $u$ is reducible in $\mathsf{Snr} \cup \mathsf{Spr}$ if and only if in the overlap graph $\gamma_u$ each nontrivial connected component (i.e., connected component containing at least one edge) contains a positive pointer.

*Example 13.9.* The micronuclear arrangement $\alpha = M_1 M_3 \overline{M}_4 M_5 M_2 M_6$ gives the realistic legals string $\varrho_6(\alpha) = v = 2\,3\,4\,\overline{5}\,\overline{4}\,5\,6\,2\,3\,6 = v_1 u_1 v_2$. Here $v$ has the legal substring $u_1 = 4\,\overline{5}\,\overline{4}\,5$, which, when removed, gives $v_1 v_2 = 2\,3\,6\,2\,3\,6$. Thus $v$ is not successful in $\mathsf{Snr} \cup \mathsf{Spr}$. Accordingly, the overlap graph $\gamma_v$, given in Fig. 13.4, has a connected component consisting of negative vertices. This component is induced by the vertices **2**, **3**, and **6**.    □

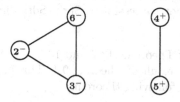

**Fig. 13.4.** The overlap graph $\gamma_v$ of Example 13.9 for the legal string $v = 2\,3\,4\,\overline{5}\,\overline{4}\,5\,6\,2\,3\,6$. The connected component containing the vertices **2**, **3**, and **6** has no positive vertices, and thus the graph has no successful reduction using the graph negative rules gnr and the graph positive rules gpr.

### 13.3.3 snr and sdr

The last case where the string negative rules are included is $\mathsf{S} = \mathsf{Snr} \cup \mathsf{Sdr}$; hence the string double rules are allowed.

**Theorem 13.3.** *A realistic legal string $u$ is successful in $\mathsf{Snr} \cup \mathsf{Sdr}$ if and only if all the pointers in $u$ are negative.*

*Proof.* The condition from the statement of the theorem is clearly necessary, since both sdr and snr reduce only negative pointers and neither of them changes the sign of any other pointers.

On the other hand, suppose that $u$ has only negative pointers. By universality of the string operations, see Theorem 9.1, there is a composition $\varphi$ of the operations $\mathsf{snr}, \mathsf{spr}, \mathsf{sdr}$ such that $\varphi(u) = \Lambda$. For all decompositions $\varphi = \varphi_2 \varphi_1$ of $\varphi$, the intermediate string $\varphi_1(u)$ contains no positive pointers (since positive pointers can be introduced only by applying spr to a positive pointer). Therefore $\varphi$ may contain only snr-rules and sdr-rules, as required by the claim.    □

Note that Theorem 13.3 translated into the language of overlap graphs states that a realistic legal string $u$ is reducible in Snr ∪ Sdr if and only if all the vertices in the overlap graph $\gamma_u$ are negative.

*Example 13.10.* (1) Consider the $\alpha$TP gene of *Sterkiella nova*, described by the micronuclear arrangement $M_1 M_3 M_5 M_7 \, M_9 M_{11} M_2 M_4 M_6 M_8 M_{10} M_{12} M_{13} M_{14}$ and thus by the realistic legal string

$$u = 2\,3\,4\,5\,6\,7\,8\,9\,10\,11\,12\,2\,3\,4\,5\,6\,7\,8\,9\,10\,11\,12\,13\,13\,14\,14.$$

By Theorem 13.3, it is successful in Snr ∪ Sdr. However, by Corollary 13.1 a rule from Snr is needed for $u$, and by Theorem 13.1 $u$ is not successful in Snr. Hence $u$ is not successful in either Snr or Sdr.

(2) The realistic legal string $2\,4\,\overline{3}\,\overline{2}\,3\,4$ corresponding to the MDS arrangement $M_1 M_4 \overline{M}_2 M_3$ is not successful in Snr ∪ Sdr, since it contains positive pointers (2 and 3).    □

As a consequence of Theorems 13.2 and 13.3, we obtain a small improvement of the universality result of Theorem 9.1. This improvement is stated as the second claim in the following theorem.

**Corollary 13.2.** *Any realistic legal string is successful in* Snr ∪ Spr ∪ Sdr. *Moreover, any elementary realistic legal string is successful in either* Snr ∪ Spr *or in* Snr ∪ Sdr.

As the next example shows the second claim of Corollary 13.2 is not true for all micronuclear genes.

*Example 13.11.* The actin gene of *Sterkiella trifallax* ([50]), is described by the micronuclear arrangement $M_3 M_4 M_6 M_5 M_7 M_9 M_{10} \overline{M}_2 M_1 M_8$ and thus by the realistic legal string

$$v = 3\,4\,4\,5\,6\,7\,5\,6\,7\,8\,9\,10\,10\,\overline{3}\,\overline{2}\,2\,8\,9.$$

By Theorem 13.2, the string $v$ is not successful in Snr ∪ Spr, because $v$ contains the legal substring $5\,6\,7\,5\,6\,7$ that has no positive pointers and it is not reducible by the string negative rules. Also, $v$ is not successful in Snr ∪ Sdr, because it has positive pointers. Nevertheless, by Theorem 9.1, $v$ is successful in Snr∪Spr∪Sdr. The same conclusions holds for the actin gene of *Sterkiella nova*, which is given by the micronuclear arrangement $M_3 M_4 M_6 M_5 M_7 M_9 \overline{M}_2 M_1 M_8$ and hence by the realistic legal string $3\,4\,4\,5\,6\,7\,5\,6\,7\,8\,9\,\overline{3}\,\overline{2}\,2\,8\,9$.    □

### 13.3.4 spr

We consider now the case where the string positive rules alone can reduce a realistic legal string to the empty string. By Lemma 13.7, disjoint cycles cannot be involved in this case. We shall first study the elementary legal strings.

**Lemma 13.9.** *An elementary realistic legal string $u$ is successful in* Spr *if and only if $u$ contains at least one positive pointer and avoids disjoint cycles.*

*Proof.* Suppose that $u$ is successful in Spr, say $\varphi_k \varphi_{k-1} \ldots \varphi_1(u) = \Lambda$, where, for each $i$, $\varphi_i = \mathrm{spr}_{p_i}$ for pointer $p_i$ (and $k = \kappa - 1$). Necessarily, $u$ has at least one positive pointer.

If $u$ contains a disjoint cycle, then, by Lemma 13.7, also the string $\varphi_i \ldots \varphi_1(u)$ contains a disjoint cycle for each $i < k$. Hence $\varphi_{k-1} \ldots \varphi_1(u)$ is a legal string of length two of the form $p\overline{p}$, but such a string does not contain a disjoint cycle. This is a contradiction.

To prove the reverse implication, assume that $u$ has at least one positive pointer and that it avoids disjoint cycles. Consider the set $S$ of positive pointers of $u$ having the maximum number of overlapping negative pointers. Furthermore, let $S'$ be that subset of $S$ consisting of pointers such that no other pointer has less overlapping positive pointers. Let $p$ be the leftmost pointer from $S'$ occurring in $u$.

If $\mathrm{spr}_p(u)$ has a legal substring containing only negative pointers, then, as in the proof of Lemma 13.8, $u$ is of the form $u = u_1 p q_1 \ldots q_i u_2 \overline{p} \overline{q}_1 \ldots \overline{q}_i u_3$, for some strings $u_1, u_2, u_3$ and pointers $q_1, q_2, \ldots, q_i$ for $i \geq 1$. In this case, however, $u$ must contain the disjoint cycle $\{pq_1, \overline{p}\overline{q}_1\}$, which is a contradiction. Therefore, any elementary substring of $\mathrm{spr}_p(u)$ has at least one positive pointer.

Assume now that $v = \mathrm{spr}_p(u)$ contains a disjoint cycle $P$. Since $u$ is realistic, there exists a realistic arrangement $\alpha$ such that $u = \varrho_\kappa(\alpha)$. Let $\delta = \psi_\kappa(\alpha)$ be the MDS descriptor of $\alpha$ such that $u = \mathrm{rem}_\kappa(\delta)$, and let $\iota$ be the IES descriptor of $\alpha$. Now $\delta$ is either of the form $\delta = \delta_1(p, q)\delta_2(\overline{p}, \overline{r})\delta_3$, or of the form $\delta = \delta_1(q, p)\delta_2(\overline{r}, \overline{p})\delta_3$ for some pointers or markers $q, r$ and MDS descriptors $\delta_1, \delta_2, \delta_3$. To make a choice, consider the former case, as the latter one can be solved similarly. Thus,

$$\delta = \delta_1(p, q)\delta_2(\overline{p}, \overline{r})\delta_3, \quad \text{and} \quad \delta' = \mathrm{spr}_p(\delta) = \delta_1 \overline{\delta}_2(\overline{q}, \overline{r})\delta_3.$$

Since $v$ contains a disjoint cycle, $\delta'$ has an IES cycle $\iota'$; let $\mathcal{I}'$ be the set of IESs occurring in $\iota'$. Since $u$ contains no disjoint cycle, $\delta$ does not have IES cycles, and so $\mathcal{I}'$ must contain some IESs of $\delta'$ that do not occur in $\delta$.

If both $\delta_1$ and $\delta_2$ are nonempty, then let $\delta_i = \delta_i'(r_i, s_i)$ for $i \in \{1, 2\}$. Also, if $\delta_3 \neq \Lambda$, then let $\delta_3 = \delta_3'(r_3, s_3)$; otherwise let $s_3 = \overline{r}$.

Next we shall distinguish three main cases depending on which of $\delta_1$ or $\delta_2$ is empty.

(i) If both $\delta_1$ and $\delta_2$ are nonempty, then the only IES of $\delta'$ that does not occur in $\delta$ is $(s_1, \overline{s}_2)$. Let $\mathcal{I}$ be the set formed by replacing $(s_1, \overline{s}_2)$ in $\mathcal{I}'$ by $(s_1, p), (s_2, \overline{p})$. However, now the IESs in $\mathcal{I}$ form an IES cycle of $\delta$; this is a contradiction.

(ii) If only one of $\delta_1, \delta_2$ is nonempty, then we have the following two subcases:

(ii.1) If $\delta_1 \neq \Lambda$, then the only IES of $\delta'$ not occurring in $\delta$ is $(s_1, \overline{q})$. Let $\mathcal{I}$ be the set formed by replacing $(s_1, \overline{q})$ in $\mathcal{I}'$ by $(s_1, p), (q, \overline{p})$. The IESs in $\mathcal{I}$ form an IES cycle of $\delta$; this is a contradiction.

(ii.2) If $\delta_2 \neq \Lambda$, then the only IES of $\delta'$ not occurring in $\delta$ is $(\overline{s}_2, s_3)$ (i.e., the connector of $\delta'$). Let $\mathcal{I}$ be the set formed by replacing $(\overline{s}_2, s_3)$ in $\mathcal{I}'$ by $(s_2, \overline{p}), (p, s_3)$. The IESs in $\mathcal{I}$ form an IES cycle of $\delta$; this is a contradiction.

(iii) If $\delta_1 = \Lambda = \delta_2$, then the only IES of $\delta'$ not occurring in $\delta$ is $(\overline{q}, s_3)$ (i.e., the connector of $\delta'$). Let $\mathcal{I}$ be the set formed by replacing $(\overline{q}, s_3)$ in $\mathcal{I}'$ by $(q, \overline{p}), (p, s_3)$. The IESs in $\mathcal{I}$ form an IES cycle of $\delta$; this is a contradiction.    □

The following lemma is an obvious consequence of Lemma 13.9.

**Lemma 13.10.** *An elementary realizable legal string $u$ is successful in* Spr *if and only if $u$ contains at least one positive pointer and avoids disjoint cycles.*

The following theorem follows directly from Lemmata 13.9 and 13.10.

**Theorem 13.4.** *A realistic string $u$ is successful in* Spr *if and only if the following holds for each legal substring $v$ of $u$: if $v = v_1 u_1 v_2 \ldots v_j u_j v_{j+1}$, where each $u_i$ is a legal substring, then $v_1 v_2 \ldots v_{j+1}$ contains at least one positive pointer and avoids disjoint cycles.*

*Proof.* The condition in the theorem is clearly necessary by Lemma 13.3. Assume then that the condition is not sufficient, and let $u$ be a realizable counterexample of minimal length. Hence $u$ is a realizable string such that for each legal substring $v$ of $u$, if $v = v_1 u_1 v_2 \ldots v_j u_j v_{j+1}$, where each $u_i$ is a legal substring, then $v_1 v_2 \ldots v_{j+1}$ contains at least one positive pointer and avoids disjoint cycles. However, $u$ is not successful in Spr. Then, by Lemma 13.9, $u$ is not elementary, and hence it contains a proper legal substring $v$, say $u = w_1 v w_2$, for some strings $w_1, w_2$. By Lemma 10.2, $v$ has a realizable conjugate $v'$. Moreover, since clearly $v$ satisfies the conditions of the theorem, so does $v'$. Indeed, if a legal string $t$ contains a disjoint cycle, then any conjugate of $t$ contains the same disjoint cycle; the same holds true with respect to positive pointers.

Since $v'$ is shorter than $u$, it follows from our assumption that $v'$ is successful in Spr, and by Lemma 13.4, so is $v$. Thus, $u$ is reduced by a composition of operations in Spr to $u' = w_1 w_2$, and so $u'$ is not successful in Spr. Since any substring of $u'$ is a scattered substring of $u$ (possibly scattered by the occurrence of $v$), it clearly follows that $u'$ satisfies the conditions of the theorem. This gives a shorter counterexample to the theorem, contradicting our assumption.    □

*Example 13.12.* (1) The MDS arrangement $M_1 M_4 \overline{M}_2 M_3$ is described by re-alistic legal string $u = 2 4 \overline{3} \overline{2} 3 4$, which is elementary. It is successful in Spr, since

$$v = 2 4 \overline{3} \overline{2} 3 4 \xrightarrow{\text{spr}_2} 3 \overline{4} 3 4 \xrightarrow{\text{spr}_{\overline{4}}} \overline{3} 3 \xrightarrow{\text{spr}_{\overline{3}}} \Lambda.$$

It is important to notice here that the string positive rules cannot be applied arbitrarily: the rule $\text{spr}_{\overline{3}}$ is applicable to $u$, yielding $\text{spr}_{\overline{3}}(u) = 2 4 2 4$, which is not successful in Spr.

(2) The realistic legal string $v = \overline{2} 2 3 4 3 4$ describing the MDS arrangement $\overline{M}_1 M_2 M_4 M_3$ is not successful in Spr, because the legal substring $3 4 3 4$ has no positive pointers.    □

### 13.3.5 sdr

In the following theorem we shall characterize the case of the realistic legal strings that can be reduced to $\Lambda$ using only the string double rules.

**Theorem 13.5.** *A realistic legal string $u$ is successful in Sdr if and only if $u$ consists of negative pointers only and it avoids disjoint cycles.*

*Proof.* Assume first that $u$ is successful in Sdr, say $\varphi_k \varphi_{k-1} \ldots \varphi_1(u) = \Lambda$, where $\varphi_i = \text{sdr}_{p_i, q_i}$ for some pointers $p_i$ and $q_i$. Clearly, $u$ consists of negative pointers only. If $u$ contains a disjoint cycle $P$, then by Lemma 13.7, so do the strings $\varphi_i \ldots \varphi_1(u)$ for each $i < k$. Then $\varphi_{k-1} \ldots \varphi_1(u)$ is a legal string of length four of the form $pqpq$, but such a string does not contain a disjoint cycle; this is a contradiction.

To prove the reverse implication, assume that $u$ avoids disjoint cycles and it has only negative pointers, but $u$ is not successful in Sdr. Suppose also that $u$ is such a string of the minimum length.

By Theorem 9.1, $u$ is successful in Snr $\cup$ Sdr, and therefore there must be a composition $\varphi$ of string double rules such that $\varphi(u)$ does have a substring $tt$ for a pointer $t$ so that $\text{snr}_t$ must be applied for $\varphi(w)$. By the minimality assumption for $u$, necessarily $\varphi = \text{sdr}_{p,q}$ for some pointers $p$ and $q$. Therefore the string $v = \text{sdr}_{p,q}(u)$ contains a disjoint cycle, say $Q$, since $v$ is realistic. Let $\delta$ and $\delta'$ be realistic descriptors such that

$$v = \text{rem}_\kappa(\delta') \quad \text{for } \delta' = \text{dlad}_{p,q}(\delta).$$

The pointers $p$ and $q$ can overlap in $\delta$ in the following ways (see p. 77):

$$\delta = \delta_1(p, r_1)\delta_2(q, r_2)\delta_3(r_3, p)\delta_4(r_4, q)\delta_5, \tag{a}$$
$$\delta = \delta_1(p, r_1)\delta_2(r_2, q)\delta_3(r_3, p)\delta_4(q, r_4)\delta_5, \tag{b}$$
$$\delta = \delta_1(r_1, p)\delta_2(q, r_2)\delta_3(p, r_3)\delta_4(r_4, q)\delta_5, \tag{c}$$
$$\delta = \delta_1(r_1, p)\delta_2(r_2, q)\delta_3(p, r_3)\delta_4(q, r_4)\delta_5, \tag{d}$$
$$\delta = \delta_1(p, r_1)\delta_2(q, p)\delta_4(r_4, q)\delta_5, \tag{e}$$
$$\delta = \delta_1(p, q)\delta_3(r_3, p)\delta_4(q, r_4)\delta_5, \tag{f}$$
$$\delta = \delta_1(r_1, p)\delta_2(q, r_2)\delta_3(p, q)\delta_5, \tag{g}$$

for some MDS descriptors $\delta_i$. In each case, one can prove that if $v$ contains a disjoint cycle, then so does $u$. We prove this in detail only for Case (c) as the other cases can be proved analogously.

In Case (c) we have

$$\delta = \delta_1(r_1, p)\delta_2(q, r_2)\delta_3(p, r_3)\delta_4(r_4, q)\delta_5,$$
$$\delta' = \delta_1(r_1, r_3)\delta_4(r_4, r_2)\delta_3\delta_2\delta_5.$$

Since $v$ contains a disjoint cycle, by Lemma 13.6, $\delta'$ has an IES cycle $\iota'$. Let $\mathfrak{I}'$ be the set of IESs of $\delta'$ occurring in $\iota'$. Since $u$ does not contain a disjoint cycle, $\delta$ does not have IES cycles, and so $\mathfrak{I}'$ must contain some IESs of $\delta'$ that do not occur in $\delta$.

If $\delta_2, \delta_3, \delta_5$ are nonempty, then let $\delta_i = (p_i, q_i)\delta_i'(p_i', q_i')$ for $i \in \{2, 3, 5\}$ (including the possibility that $\delta_i' = \Lambda$ and $\delta_i = (p_i, q_i) = (p_i', q_i')$). Also, if $\delta_1 \neq \Lambda$, then let $\delta_1 = (p_1, q_1)\delta_1'$; otherwise let $p_1 = r_1$.

We distinguish the following four main cases, depending on how many of $\delta_2, \delta_3, \delta_5$ are empty MDS descriptors. In each of these cases we can build a set $\mathfrak{I}$ such that the IESs from $\mathfrak{I}$ form an IES cycle of $\delta$, leading to contradiction.

(i) If $\delta_2, \delta_3, \delta_5$ are all nonempty, then $(q_3', p_2)$ and $(q_2', p_5)$ are the only IESs of $\delta'$ not occurring in $\delta$. If $(q_3', p_2) \in \mathfrak{I}'$, then let $\mathfrak{I} = (\mathfrak{I}' \setminus \{(q_3', p_2)\}) \cup \{(q_3', p), (p, p_2)\}$. If $(q_2', p_5) \in \mathfrak{I}'$, then let $\mathfrak{I} = (\mathfrak{I}' \setminus \{(q_2', p_5)\}) \cup \{(q_2', q), (q, p_5)\}$. In both cases the IESs in $\mathfrak{I}$ form an IES cycle of $\delta$, which is a contradiction.

(ii) If only one of $\delta_2, \delta_3, \delta_5$ is empty, then we have the following subcases:

(ii.1) If $\delta_2 = \Lambda$, then the IES occurring in $\delta'$ but not in $\delta$ is $(q_3', p_5)$. Then $\mathfrak{I} = (\mathfrak{I}' \setminus \{(q_3', p_5)\}) \cup \{(q_3', p), (p, q), (q, p_5)\}$. Observe that $(p, q)$ is indeed an IES of $\delta$ in this case.

(ii.2) If $\delta_3 = \Lambda$, then the IESs occurring in $\delta'$ but not in $\delta$ are $(r_2, p_2)$ and $(q_2', p_5)$. If $(r_2, p_2) \in \mathfrak{I}'$, then let $\mathfrak{I} = (\mathfrak{I}' \setminus \{(r_2, p_2)\}) \cup \{(r_2, p), (p, p_2)\}$. If $(q_2', p_5) \in \mathfrak{I}'$, then let $\mathfrak{I} = (\mathfrak{I}' \setminus \{(q_2', p_5)\}) \cup \{(q_2', q), (q, p_5)\}$.

(ii.3) If $\delta_5 = \Lambda$, then the IESs occurring in $\delta'$ but not in $\delta$ are $(q_3', p_2)$ and $(q_2', p_1)$. If $(q_3', p_2) \in \mathfrak{I}'$, then let $\mathfrak{I} = (\mathfrak{I}' \setminus \{(q_3', p_2)\}) \cup \{(q_3', p), (p, p_2)\}$. If $(q_2', p_1) \in \mathfrak{I}'$, then let $\mathfrak{I} = (\mathfrak{I}' \setminus \{(q_2', p_1)\}) \cup \{(q_2', q), (q, p_1)\}$.

(iii) If two of $\delta_2, \delta_3, \delta_5$ are empty, then we have the following subcases:

(iii.1) If $\delta_2 = \delta_3 = \Lambda$, then the IES occurring in $\delta'$ but not in $\delta$ is $(r_2, p_5)$. Then $\mathfrak{I} = (\mathfrak{I}' \setminus \{(r_2, p_5)\}) \cup \{(r_2, p), (p, q), (q, p_5)\}$.

(iii.2) If $\delta_2 = \delta_5 = \Lambda$, then the IES occurring in $\delta'$ but not in $\delta$ is $(q_3', p_1)$. Then $\mathfrak{I} = (\mathfrak{I}' \setminus \{(q_3', p_1)\}) \cup \{(q_3', p), (p, q), (q, p_1)\}$.

(iii.3) If $\delta_3 = \delta_5 = \Lambda$, then the IESs occurring in $\delta'$ but not in $\delta$ are $(r_2, p_2)$ and $(q_2', p_1)$. If $(r_2, p_2) \in \mathfrak{I}'$, then let $\mathfrak{I} = (\mathfrak{I}' \setminus \{(r_2, p_2)\}) \cup \{(r_2, p), (p, p_2)\}$. If $(q_2', p_1) \in \mathfrak{I}'$, then let $\mathfrak{I} = (\mathfrak{I}' \setminus \{(q_2', p_1)\}) \cup \{(q_2', q), (q, p_1)\}$.

(iv) If $\delta_2 = \delta_3 = \delta_5 = \Lambda$, then the only IES of $\delta'$ not occurring in $\delta$ is $(r_2, p_1)$ and $\mathfrak{I} = (\mathfrak{I}' \setminus \{(r_2, p_1)\}) \cup \{(r_2, p), (p, q), (q, p_1)\}$.

Consequently, $v$ does not contain a disjoint cycle and thus, since $v$ is shorter than $u$, $v$ must be successful in Sdr. However, since $u = \mathsf{sdr}_{p,q}(u)$, it follows that $u$ is also successful in Sdr; this is a contradiction.     □

*Example 13.13.* According to Theorem 13.5, the realistic legal string $2\,3\,2\,3$ corresponding to the micronuclear arrangement $M_2 M_1 M_3$ is successful in Sdr, while the realistic legal string $2\,2\,3\,3$ corresponding to $M_1 M_2 M_3$ is not successful in Sdr.     □

### 13.3.6 spr and sdr

The case where we allow both string positive and string double rules is now clear, since by using Theorems 13.5 and 13.4, we have

**Theorem 13.6.** *A realistic legal string $u$ is successful in* $\mathsf{Spr} \cup \mathsf{Sdr}$ *if and only if $u$ avoids disjoint cycles.*

Theorem 13.6 has the following corollary.

**Corollary 13.3.** *An elementary realistic legal string $u$ is successful in* $\mathsf{Spr} \cup \mathsf{Sdr}$ *if and only if either $u$ is successful in* Spr *or in* Sdr.

*Example 13.14.* Consider the micronuclear arrangement $M_1 \overline{M}_2 M_4 M_3$. The corresponding elementary realistic legal string is then $u = 2\,\overline{3}\,\overline{2}\,4\,3\,4$. Then $u$ is successful in $\mathsf{Spr} \cup \mathsf{Sdr}$, since it has no disjoint cycles. Indeed, we have the following successful string reduction for $u$:

$$2\,\overline{3}\,\overline{2}\,4\,3\,4 \xrightarrow{\ \mathsf{spr}_2\ } 3\,4\,3\,4 \xrightarrow{\ \mathsf{sdr}_{3,4}\ } \Lambda.$$

By Corollary 13.3, $u$ is also successful either in Spr or in Sdr. Since $u$ has positive pointers, the former case must hold. As a matter of fact, the only successful reduction for $u$ in Spr is:

$$2\,\overline{3}\,\overline{2}\,4\,3\,4 \xrightarrow{\ \mathsf{spr}_{\overline{3}}\ } 2\,\overline{4}\,2\,4 \xrightarrow{\ \mathsf{spr}_{\overline{4}}\ } 2\,\overline{2} \xrightarrow{\ \mathsf{spr}_2\ } \Lambda.$$

□

## 13.4 Complexity of Reductions

One can use the results of this chapter to set a measure of similarity for micronuclear genes and also to classify these genes by their complexity. For instance, two seemingly different micronuclear genes may be considered similar if both of them can be assembled using only string positive rules. Also, the genes that need all three types of operations in any successful assembly strategy may be considered more complex than the genes needing only one or two of these types. In this sense, the micronuclear actin gene of *Sterkiella nova* (see Example 8.4, or Example 13.15, below) is more complex than the micronuclear $\alpha$TP gene of *Sterkiella nova* (see Example 13.10(1)). As a matter of fact, one can use any order of the complexity of the three operations (say, snr $<$ spr $<$ sdr) to obtain a comparative classification (a partial order) of the complexity of all micronuclear genes. Such an order of the complexity of the three operations may be motivated by either some biological or mathematical considerations.

One can also measure complexity of specific reductions. In a rather simplified approach, the operations snr, spr, and sdr can be assigned different complexities, say $c_{snr}$, $c_{spr}$, and $c_{sdr}$, so that the complexity of a single application of $spr_p$ is measured by the constant $c_{spr}$. Then a reduction $\varphi(v) = \Lambda$ has the complexity

$$c(v) = n_\ell \cdot c_{snr} + n_h \cdot c_{spr} + n_d \cdot c_{sdr},$$

where $n_\ell$, $n_h$ and $n_d$ denote the number of the applications of the corresponding operations in the reduction.

*Example 13.15.* Consider the actin gene in *Sterkiella nova* that is described by the realistic legal string $v = 3\,4\,4\,5\,6\,7\,5\,6\,7\,8\,9\,\overline{3}\,\overline{2}\,2\,8\,9$. The string has the following reduction to the empty string:

$$v = 3\,4\,4\,5\,6\,7\,5\,6\,7\,8\,9\,\overline{3}\,\overline{2}\,2\,8\,9 \xrightarrow{\text{snr}_4} 3\,5\,6\,7\,5\,6\,7\,8\,9\,\overline{3}\,\overline{2}\,2\,8\,9$$

$$\xrightarrow{\text{spr}_{\overline{2}}} 3\,5\,6\,7\,5\,6\,7\,8\,9\,\overline{3}\,8\,9 \xrightarrow{\text{spr}_3} 9\,\overline{8}\,7\,\overline{6}\,5\,7\,\overline{6}\,5\,8\,9$$

$$\xrightarrow{\text{spr}_{\overline{9}}} \overline{8}\,5\,6\,7\,5\,6\,7\,8 \xrightarrow{\text{spr}_{\overline{8}}} \overline{7}\,6\,5\,\overline{7}\,6\,5$$

$$\xrightarrow{\text{sdr}_{\overline{7},\overline{6}}} \overline{5}\,\overline{5} \xrightarrow{\text{snr}_{\overline{5}}} \Lambda.$$

Then the complexity of this reduction is $2 \cdot c_{snr} + 4 \cdot c_{spr} + 1 \cdot c_{sdr}$. Here is another reduction for $v$:

$$v = 3\,4\,4\,5\,6\,7\,5\,6\,7\,8\,9\,\overline{3}\,\overline{2}\,2\,8\,9 \xrightarrow{\text{snr}_4} 3\,5\,6\,7\,5\,6\,7\,8\,9\,\overline{3}\,\overline{2}\,2\,8\,9$$

$$\xrightarrow{\text{spr}_{\overline{2}}} 3\,5\,6\,7\,5\,6\,7\,8\,9\,\overline{3}\,8\,9 \xrightarrow{\text{sdr}_{5,6}} 3\,7\,7\,8\,9\,\overline{3}\,8\,9$$

$$\xrightarrow{\text{sdr}_{8,9}} 3\,7\,7\,\overline{3} \xrightarrow{\text{snr}_7} 3\,\overline{3} \xrightarrow{\text{spr}_3} \Lambda.$$

The complexity of this reduction is $2 \cdot c_{snr} + 2 \cdot c_{spr} + 2 \cdot c_{sdr}$. The latter reduction is easier, if only operations are counted. The former reduction uses seven and the latter uses only six operations. On the other hand, the latter reduction is harder, if $c_{sdr} > 2 \cdot c_{spr}$.    □

In the above we compared different reductions for legal strings (and thus for reassembling micronuclear genes). The complexity measures in these considerations did not take into account the distances between the applied occurrences of pointers. For instance, the operation $spr_p$ may have different complexity for different strings; one could define the complexity depending on how many IESs lie between the two occurrences of pointers from $\{p, \overline{p}\}$, see the discussion on the simple variants of our operations in Sect. 16.3.


## Notes on References

- The results of this chapter are from Ehrenfeucht et al. [16, 17].
- The graph theoretic counterpart of Theorem 13.2 mentioned on p. 141 was proved independently by Anderson et al. [3].
- The graph theoretic characterizations for the cases where graph negative rules gnr are used are rather simple as can be seen in the corresponding subsections. When we admit only gpr and gdr, the characterization problems are challenging and they are currently open.
- The commutation results, Lemmata 13.1 and 13.2, are new. These results have been designed to simplify the proof of Lemma 13.3 of [16, 17]. Also, Lemma 13.7 is new and is stated to emphasize the impact of the string positive rules and the string double rules on the disjoint cycles.
- Complexity issues for the operations snr, spr, and sdr are considered by Daley [8] from the standard algorithmic and the more novel "biocomplexity" points of view.

# Gene Assembly Through Cyclic Graph Decomposition

In this chapter we consider a graph theoretic framework for gene assembly where segments of genes are distributed over a set of circular molecules. Our main motivation still lies in the gene assembly process of ciliates. However, the model in this chapter is more general, and it can be seen as a graph theoretic formulation of the "fold and recombine" computing paradigm.

## 14.1 Graphs with Labels and Colors

We shall consider (multi)graphs, which will be called MI-graphs, where the intuition comes from MDS/IES structures as considered in the previous chapters. They have two kinds of edge labels, where one kind is called colors to avoid ambiguity. In this chapter, we work with directed edges in the sense that each undirected edge $\{x, y\}$ between two vertices is oriented in both directions, one going from $x$ to $y$, and the other going from $y$ to $x$.

Let $V$ be a finite set, and let $V \times V = \{(x, y) \mid x, y \in V\}$ be the set of all ordered pairs of elements of $V$. For each pair $a = (x, y) \in V \times V$, let $\bar{a} = (y, x)$ be the **reverse** of $a$.

An **MI-graph** is a 5-tuple $\gamma = (V, E, \varepsilon, f, h)$, where

- $V$ is a finite set of **vertices**
- $E$ is a signed alphabet of **edges**
- $\varepsilon\colon E \to V \times V$ is an **endpoint mapping**,
- $f\colon E^{\circledast} \to \Gamma^{\circledast}$ is a morphism, called the **labelling function**, for some signed alphabet $\Gamma$,
- $h\colon E \to \{1, 2\}$ is a **coloring function**,

and the following conditions are satisfied: for all $e \in E$,

$$e \in E \iff \bar{e} \in E, \tag{14.1}$$

$$\varepsilon(\bar{e}) = \overline{\varepsilon(e)}, \quad f(\bar{e}) = \overline{f(e)}, \quad \text{and} \quad h(e) = h(\bar{e}). \tag{14.2}$$

The values 1 and 2 of the coloring function $h$ are referred to as **colors**. Also, recall that $\overline{\overline{e}} = e$ for all $e \in E$.

An MI-graph $\gamma$ can be thought to describe a folded DNA molecule divided into MDS and IES regions. The labelling function $f$ of $\gamma$ attaches a string and its inversion to the two orientations of the edges – such a string (label) represents a submolecule of the DNA molecule represented by the graph. On the other hand, the colors 1 and 2 of the edges indicate MDS and IES regions of the whole micronuclear molecule.

For an edge $e \in E$ with $\varepsilon(e) = (x, y)$, $x$ is the **initial vertex** of $e$, denoted by $\iota(e)$, and $y$ is the **terminal vertex** of $e$, denoted by $\tau(e)$; hence $\varepsilon(e) = (\iota(e), \tau(e))$. Moreover, the vertices $\iota(e)$ and $\tau(e)$ are the **ends** of $e$. An edge $e \in E$ with $\iota(e) = \tau(e)$ is a **loop**; note that for a loop $e$, $\varepsilon(e) = \varepsilon(\overline{e})$, but always $e \neq \overline{e}$. In general, we say that two edges $e, e' \in E$ are **parallel** if $\varepsilon(e) = \varepsilon(e')$.

If needed, the components of an MI-graph are identified by subscripts, i.e.,

$$\gamma = (V_\gamma, E_\gamma, \varepsilon_\gamma, f_\gamma, h_\gamma).$$

For a vertex $x \in V_\gamma$, let

$$E_\gamma^+(x) = \{e \in E_\gamma \mid \tau(e) = x\} \quad \text{and} \quad E_\gamma^-(x) = \{e \in E_\gamma \mid \iota(e) = x\}$$

be the set of edges with $x$ as the initial and terminal vertex, respectively. The **valency** of a vertex $x$, denoted by $\mathrm{val}_\gamma(x)$, is the number of edges with $x$ as the terminal vertex, that is,

$$\mathrm{val}_\gamma(x) = \mathrm{card}(E_\gamma^+(x)).$$

By condition (14.1), $\mathrm{val}_\gamma(x)$ is also equal to the number of edges with $x$ as the initial vertex. Notice that if $\varepsilon_\gamma(e) = (x, x)$, then both $e$ and $\overline{e}$ are in $E_\gamma^+(x)$ (and also in $E_\gamma^-(x)$).

For each vertex $x \in V_\gamma$ and each color $c \in \{1, 2\}$, let

$$\mathrm{val}_\gamma(x; c) = \mathrm{card}(\{e \in E_\gamma^+(x) \mid h_\gamma(e) = c\})$$

be the number of edges of color $c$ with $x$ as the terminal vertex. Now, $\mathrm{val}_\gamma(x) = \mathrm{val}_\gamma(x; 1) + \mathrm{val}_\gamma(x; 2)$. A vertex $x \in V_\gamma$ is **balanced** if $\mathrm{val}_\gamma(x; 1) = \mathrm{val}_\gamma(x; 2)$; then $\gamma$ is **balanced** if every vertex $x \in V_\gamma$ is balanced.

We adopt now some conventions for *drawing MI-graphs*. Let then $\gamma$ be an MI-graph. We shall draw the edges $e \in E_\gamma$ colored by 1 (i.e., $h_\gamma(e) = 1$) using solid lines with arrows and those colored by 2 using dashed lines with arrows. Also, to simplify the drawings, we shall draw only one of the two edges that are reverses of each other – clearly, we still have an unambiguous representation of $\gamma$, independently of which direction we choose to be drawn. As usual, the label $f_\gamma(e)$ of a drawn edge $e$ is given next to the drawing of $e$.

*Example 14.1.* Let $\Gamma = \{a\}$ be the alphabet of edge labels, and let $b = \bar{a}$. Also, let $\gamma$ be the MI-graph given in Fig. 14.1a, where $V_\gamma = [1, 8]$. Here $h_\gamma(e) = 1$ for the solid edges (and their reversals). The labels can be read from the edges: e.g., $f_\gamma(e) = abb$ and $f_\gamma(\bar{e}) = aab$ $(= \overline{abb})$ for the edge $e$ with $\varepsilon(e) = (7, 8)$. The MI-graph $\gamma$ is balanced; e.g., $\mathrm{val}_\gamma(1) = 4$ and $\mathrm{val}(1; 1) = 2 = \mathrm{val}(1; 2)$.

In Fig. 14.1b the same graph $\gamma$ is drawn differently by choosing to drawn the reverses of some of the edges. $\qquad\square$

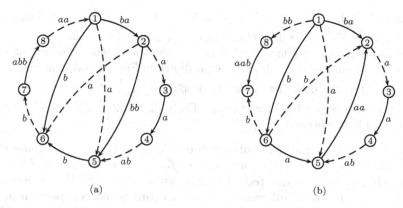

(a)　　　　　　　　　　(b)

**Fig. 14.1. a,b.** The MI-graph $\gamma$ of Example 14.1 has been given two different representations in **a** and **b** by choosing a different orientation for some of the edges. The colors (*solid/dashed lines*) of the edges remain the same in both representations, but the labels of the reversed edges are inverses of each other.

A **walk** in an MI-graph $\gamma$ is a string $\pi = e_1 e_2 \ldots e_n$ over the alphabet $E_\gamma$ such that $\tau(e_i) = \iota(e_{i+1})$ for each $i \in [1, n-1]$. The vertex $\iota(e_1)$ is the **initial vertex** of the walk $\pi$, and it is denoted by $\iota(\pi)$. Similarly, the last vertex $\tau(e_n)$ is the **terminal vertex** of $\pi$, and it is denoted by $\tau(\pi)$. Also, the initial and the terminal vertices of $\pi$ are called the **ends** of the walk $\pi$, and we write $\pi : \iota(\pi) \to \tau(\pi)$, or with some ambiguity (for MI-graphs with parallel edges) we present such a walk as a sequence

$$\pi : \quad x_1 \xrightarrow{c_1} x_2 \xrightarrow{c_2} \ldots \xrightarrow{c_n} x_{n+1},$$

where $\varepsilon_\gamma(e_i) = (x_i, x_{i+1})$ and $c_i = h_\gamma(e_i)$. We say that each vertex $x_i$, for $i \in [1, n+1]$, and each edge $e_j$, for $j \in [1, n]$, is **on the walk** $\pi$. The walk $\pi$ is **closed** if $x_1 = x_{n+1}$, i.e., $\iota(\pi) = \tau(\pi)$. The **label** of the walk $\pi$ is the string

$$f_\gamma(\pi) = f_\gamma(e_1) f_\gamma(e_2) \ldots f_\gamma(e_n),$$

obtained by concatenating the labels of its edges. For a walk $\pi = e_1 e_2 \ldots e_n$, the walk $\bar{\pi} = \bar{e}_n \bar{e}_{n-1} \ldots \bar{e}_1$ is the **reverse walk** of $\pi$.

Lemma 14.1 follows from the definitions.

**Lemma 14.1.** *Let $\pi$ be a walk in an MI-graph $\gamma$. The label of the reverse walk $\overline{\pi}$ is the inversion of the label of $\pi$, that is, $f_\gamma(\overline{\pi}) = \overline{f_\gamma(\pi)}$.*

If a walk $\pi_1$ terminates where a walk $\pi_2$ starts, that is, $\tau(\pi_1) = \iota(\pi_2)$, then $\pi_1\pi_2\colon \iota(\pi_1) \to \tau(\pi_2)$ is their **composed walk** where $\pi_1\pi_2$ is the string concatenation of these walks. Clearly, $f_\gamma(\pi_1\pi_2) = f_\gamma(\pi_1)f_\gamma(\pi_2)$.

Let $\pi = e_1e_2\ldots e_n$ be a walk, where $\varepsilon_\gamma(e_i) = (x_i, x_{i+1})$ for each $i$. Then $\pi$ is a **path** if it never enters the same vertex twice: $x_i \neq x_j$ for all $i \neq j$. Also, the walk $\pi$ is a **cycle** if it is closed and $x_i \neq x_j$ for all $i \neq j$ with $i, j \leq n$.

*Example 14.2.* Let $\gamma$ be the MI-graph of Example 14.1 (Fig. 14.1a). The walk $\pi\colon 1 \xrightarrow{1} 2 \xrightarrow{1} 5 \xrightarrow{1} 6 \xrightarrow{1} 1$ is closed, and $h_\gamma(e) = 1$ (solid) for all edges in this walk. Moreover, we have that $f_\gamma(\pi) = babbb\overline{b} = babbba$. Notice that the reverse of the edge $e$ with $\varepsilon_\gamma(e) = (6, 1)$ is drawn in Fig. 14.1(a), and therefore $f_\gamma(e) = a$. On the other hand, the walk $\pi'\colon 1 \xrightarrow{1} 2 \xrightarrow{1} 5 \xrightarrow{0} 1 \xrightarrow{0} 8$ is not a path, since the vertex 1 is visited twice. The walk $\pi''\colon 2 \xrightarrow{0} 3 \xrightarrow{1} 4 \xrightarrow{0} 5 \xrightarrow{1} 2$ is closed, and it is a cycle in $\gamma$.    □

An MI-graph $\gamma'$ is an **MI-subgraph** of $\gamma$ if $V_{\gamma'} \subseteq V_\gamma$, $E_{\gamma'} \subseteq E_\gamma$, and the functions $f_{\gamma'}$ and $h_{\gamma'}$ are the restrictions of $f_\gamma$ and $h_\gamma$ on the edges of $\gamma'$.

An MI-graph $\gamma$ is **connected**, if for any two vertices $x, y \in V_\gamma$, there exists a walk $\pi\colon x \to y$. An MI-subgraph $\gamma'$ is called a **connected component** of $\gamma$, if $\gamma'$ is a maximal connected MI-subgraph of $\gamma$ (that is, if $\gamma'$ is not an MI-subgraph of another connected MI-subgraph of $\gamma$).

*Example 14.3.* Consider the MI-graph $\gamma$ given in Example 14.1, which has two drawings in Fig. 14.1. An MI-subgraph $\gamma'$ of $\gamma$, given in Fig. 14.2, has the vertex set $V' = V \setminus \{8\}$, and the edge set $E'$ is a subset of the edge set $E$ of $\gamma$. The MI-graph $\gamma'$ has two connected components: the first one $\gamma_1$ has the vertex set $V_1 = \{1, 2, 6, 7\}$ (with the edges of $\gamma'$ that have both ends in $V_1$), and the second one $\gamma_2$ has the vertex set $V_2 = \{3, 4, 5\}$ (with the two edges of $\gamma'$ that have both ends in $V_2$).    □

We say that $\gamma$ is a **cyclic MI-graph** if all its vertices and edges are on one cycle. For disjoint MI-graphs $\gamma_1, \ldots, \gamma_m$ (i.e., with $V_{\gamma_i} \cap V_{\gamma_j} = \emptyset$ for $i \neq j$), their **disjoint union** denoted by

$$\gamma = \sum_{i=1}^{m} \gamma_i,$$

is the MI-graph with $V_\gamma = \cup_{i=1}^{m} V_{\gamma_i}$ and $E_\gamma = \cup_{i=1}^{m} E_{\gamma_i}$ such that each $\gamma_i$ is a connected component of $\gamma$.

Two MI-graphs $\gamma$ and $\gamma'$ are said to be **isomorphic** if there are bijections $\varphi\colon V_\gamma \to V_{\gamma'}$ and $\rho\colon E_\gamma \to E_{\gamma'}$ such that for all edges $e \in E_\gamma$ with $\varepsilon_\gamma(e) = (x, y)$, we have

$$\varepsilon_{\gamma'}(\rho(e)) = (\varphi(x), \varphi(y)), \quad f_{\gamma'}(\rho(e)) = f_\gamma(e), \quad \text{and} \quad g_{\gamma'}(\rho(e)) = g_\gamma(e).$$

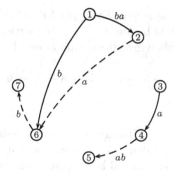

**Fig. 14.2.** The MI-subgraph $\gamma'$ of the MI-graph $\gamma$ in Example 14.1 (see Fig. 14.1) has the vertex set $V' = [1,7]$. The vertices $1, 2, 6, 7$ (and the connecting edges of $\gamma'$) form a connected component of $\gamma'$. Also, the vertices $\{3, 4, 5\}$ form a connected component of $\gamma'$.

*Remark 14.1.* The notion of isomorphism requires that the corresponding edges, $e$ and $\rho(e)$, have *identical* labels and colors. In this chapter *we shall identify isomorphic MI-graphs*: if the MI-graphs $\gamma$ and $\gamma'$ are isomorphic, then we shall simply write $\gamma = \gamma'$. □

Let $\pi = e_1 e_2 \ldots e_n$ be a walk in an MI-graph $\gamma$. Then $\pi$ is called

- a **trail** if each edge is visited at most once in either direction: $e_i \neq e_j$ and $e_i \neq \bar{e}_j$ for all $i \neq j$;
- a **maximal trail** if it is a trail and there exists no edge $e \in E_\gamma$ such that $e_1 e_2 \ldots e_n e$ is a trail; we also say then that $\pi$ is a **maximal trail of** $e_1$;
- an **Euler trail** if it is a trail that contains each edge (in one orientation): for each $e \in E_\gamma$, there is an index $i$ such that either $e = e_i$ or $\bar{e} = e_i$;
- an **alternating** walk if $h_\gamma(e_i) \neq h_\gamma(e_{i+1})$ for all $i \in [1, n-1]$;
- an **alternating closed walk** if $\pi$ is alternating, closed, and, furthermore, $h_\gamma(e_n) \neq h_\gamma(e_1)$.

An MI-graph $\gamma$ is said to be (**alternating**) **Eulerian** if $\gamma$ has an (alternating) closed Euler trail, and $\gamma$ is **even** if the valency $\mathrm{val}_\gamma(x)$ is even for each $x \in V_\gamma$.

The following result is known as **Euler's theorem.**

**Theorem 14.1.** *An MI-graph $\gamma$ is Eulerian if and only if it is connected and even.*

The following basic result is used in later sections.

**Theorem 14.2.** *An even MI-graph $\gamma$ is alternating Eulerian if and only if $\gamma$ is connected and balanced.*

An MI-graph $\gamma$ is a **recombination MI-graph** if for each vertex $x$, $\mathrm{val}_\gamma(x) = 2$ or $4$, and every vertex of valency 4 is balanced, i.e., $\mathrm{val}_\gamma(x; 1) = 2 = \mathrm{val}_\gamma(x; 2)$.

*Example 14.4.* The MI-graph $\gamma$ in Example 14.1 (Fig. 14.1) is a recombination MI-graph. Indeed, $\gamma$ is balanced, since the vertices of valency 2 are also balanced. We have the following alternating closed Euler trail

$$1 \xrightarrow{1} 2 \xrightarrow{2} 3 \xrightarrow{1} 4 \xrightarrow{2} 5 \xrightarrow{1} 6 \xrightarrow{2} 7 \xrightarrow{1} 8 \xrightarrow{2} 1 \xrightarrow{1} 6 \xrightarrow{2} 2 \xrightarrow{1} 5 \xrightarrow{2} 1$$

in $\gamma$.                                                                           □

Note that recombination MI-graphs are not necessarily alternating Eulerian, since the vertices of valency 2 need not be balanced. However, by the following lemma, they have a closed Euler trail $\pi$ that **alternates at valency 4 vertices** (i.e., $\pi = e_1 e_2 \ldots e_n$, with $\varepsilon_\gamma(e_i) = (x_i, x_{i+1})$ and $x_{n+1} = x_1$, such that for all $i$, $h_\gamma(e_i) \neq h_\gamma(e_{i+1})$ if $\mathrm{val}_\gamma(x_{i+1}) = 4$).

**Lemma 14.2.** *A connected recombination MI-graph $\gamma$ has a closed Euler trail that alternates at valency 4 vertices.*

*Proof.* If $\gamma$ is a cyclic MI-graph, then the claim is trivial, since, in this case, all vertices have valency 2. Assume then that $\gamma$ is not a cyclic MI-graph. Recall that the unbalanced vertices $x \in V_\gamma$ have valency 2, and the two edges in $E_\gamma^+(x)$ have the same color. Define a new MI-graph $\gamma'$ as follows: for each unbalanced vertex $x \in V_\gamma$ with $\mathrm{val}_\gamma(x; c) = 2$ (where either $c = 1$ or $c = 2$), add a loop $\varepsilon_\gamma(e) = (x, x)$ with $h_{\gamma'}(e) = 3 - c$ and $f_{\gamma'}(e) = \Lambda$. The MI-graph $\gamma'$ is balanced, and therefore it has an alternating closed Euler trail $\pi'$ by Theorem 14.2. Clearly, the corresponding path $\pi$ with the introduced loops removed satisfies the claim in the original recombination MI-graph $\gamma$.

                                                                           □

## 14.2 Folding an MI-graph

Let $\gamma$ be an even MI-graph (i.e., with even valencies). A **pairing function** $\chi$ of $\gamma$ chooses for each edge $e \in E_\gamma$ incoming to a vertex $x$ (i.e., $e \in E_\gamma^+(x)$) an edge outgoing from $x$ (i.e., $\chi(e) \in E_\gamma^-(x)$) such that

$$\chi(\overline{\chi(e)}) = \overline{e}, \tag{14.3}$$

$$\chi(e) = \overline{e} \iff e \text{ is a loop.} \tag{14.4}$$

By condition (14.3), if $\chi$ chooses $e_2$ for $e_1$ ($\xrightarrow{e_1} x \xrightarrow{e_2}$) then it chooses the reverse edge of $e_1$ for the reverse of $e_2$ ($\xleftarrow{\overline{e_1}} x \xleftarrow{\overline{e_2}}$). By condition (14.4), the pairing does not choose the reverse of the incoming edge unless the edge is a loop. Clearly, a pairing function is a permutation of the edge set $E_\gamma$.

Let $\chi$ be a pairing function of an even MI-graph $\gamma$. For each edge $e_1 \in E_\gamma$, let $\pi_\chi(e_1) = e_1 e_2 \ldots e_k$ be the maximal trail of $e_1$ uniquely obtained by following $\chi$: $\chi(e_{i-1}) = e_i$ for all $i \in [2, k]$. Since $\chi$ is a bijection, the trail $\pi_\chi(e)$

is well defined for each $e \in E_\gamma$. Note also that if $e \in E_\gamma$ is a loop, then we have simply that $\pi_\chi(e) = e$.

For a recombination MI-graph $\gamma$, the **natural pairing function** $\eta_\gamma$ is defined so that it pairs the same colored edges for vertices of valency 4 and the only possible edges for the vertices $x$ of valency 2. Formally, if $e \in E_\gamma^+(x)$, $e' \in E_\gamma^-(x)$ with $\bar{e} \neq e'$, then

$$\eta_\gamma(e) = e' \iff \text{either } \text{val}_\gamma(x) = 2, \text{ or } \text{val}_\gamma(x) = 4 \text{ and } h_\gamma(e) = h_\gamma(e').$$

We shall usually omit the index from $\eta_\gamma$, and simply write $\eta$ for the natural pairing function.

*Example 14.5.* Consider the recombination MI-graph $\gamma$ in Example 14.1 (see Fig. 14.1). Now, for instance, $\eta(e_1) = e_2$ for $\varepsilon_\gamma(e_1) = (8,1)$ and $\varepsilon_\gamma(e_2) = (1,5)$ (and also $\chi(\bar{e}_2) = \bar{e}_1$). Also, $\eta(e_3) = e_4$ for $\varepsilon_\gamma(e_3) = (2,1)$ and $\varepsilon_\gamma(e_4) = (1,6)$, and $\eta(e_5) = e_6$ for $\varepsilon_\gamma(e_5) = (2,3)$ and $\varepsilon_\gamma(e_6) = (3,4)$. The pairing function $\eta$ gives a partition of the edges into the following closed trails (and their reversals): $\pi_1 \colon 1 \xrightarrow{1} 2 \xrightarrow{1} 5 \xrightarrow{1} 6 \xrightarrow{1} 1$ and $\pi_2 \colon 1 \xrightarrow{2} 5 \xrightarrow{2} 4 \xrightarrow{1} 3 \xrightarrow{2} 2 \xrightarrow{2} 6 \xrightarrow{2} 7 \xrightarrow{1} 8 \xrightarrow{2} 1$. $\quad\square$

The following result shows that maximal trails are closed.

**Theorem 14.3.** *Let $\chi$ be a pairing function of an even MI-graph $\gamma$. Then the maximal trail $\pi_\chi(e)$ is closed for each $e \in E_\gamma$, and the edge sets of the maximal trails $\pi_\chi(e)$, for $e \in E_\gamma$, form a partition of $E_\gamma$.*

In the graph theoretic framework of this chapter, the process of gene assembly in ciliates is represented by a two-stage processing: first, folding the circular MI-graphs (corresponding to circular DNA molecules), and, second, unfolding the folded MI-graphs by splitting the vertices of valency 4.

Following this intuition, for an MI-graph $\gamma$, a $\gamma$-**pointer** is a pair $p = \{x, y\}$, where $x$ and $y$ are different vertices. The vertices $x$ and $y$ are called the **ends** of the $\gamma$-pointer $p$. A set $P$ of mutually disjoint $\gamma$-pointers is called a $\gamma$-**pointer family**.

Let $p = \{x, y\}$ be a $\gamma$-pointer, and let $V = (V_\gamma \setminus \{x, y\}) \cup \{p\}$, where we assume that $p \notin V_\gamma$. The $p$-**folded MI-graph** of $\gamma$ is the MI-graph

$$\gamma * p = (V, E_\gamma, \varepsilon, f_\gamma, h_\gamma),$$

obtained by identifying the ends of $p$: for each $e \in E_\gamma$ with $\varepsilon_\gamma(e) = (u, v)$,

$$\varepsilon(e) = (u', v'), \text{ where } u' = \begin{cases} u, & \text{if } u \notin p, \\ p, & \text{if } u \in p, \end{cases} \text{ and } v' = \begin{cases} v, & \text{if } v \notin p, \\ p, & \text{if } v \in p. \end{cases}$$

We illustrate the folding operation in Fig. 14.3.

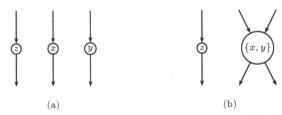

(a)                                          (b)

**Fig. 14.3. a,b.** Folding with respect to the $\gamma$-pointer $p = \{x, y\}$, where $x$ and $y$ are different vertices of the MI-graph. In **a** we have sketched the situation in $\gamma$ for a vertex $z$ not in the $\gamma$-pointer $p$, and with respect to the vertices $x$ and $y$. In **b** the folding at the $\gamma$-pointer $p$ is illustrated.

Note that the folded MI-graph $\gamma * p$ has the same edge set and the same labelling and coloring functions as the original MI-graph $\gamma$. However, the folded MI-graph has fewer vertices: $\mathrm{card}(V_{\gamma*p}) = \mathrm{card}(V_\gamma) - 1$. Also, some edges may become parallel in $\gamma * p$ even though they were not parallel in $\gamma$. Indeed, if $p = \{x, y\}$ and $e_1, e_2 \in E_\gamma$ are such that $\varepsilon_\gamma(e_1) = (x, z)$ and $\varepsilon_\gamma(e_2) = (y, z)$, then $\varepsilon_{\gamma*p}(e_1) = (p, z) = \varepsilon_{\gamma*p}(e_2)$.

*Example 14.6.* Let $\gamma$ be the recombination MI-graph from Fig. 14.1, and let $p = \{1, 7\}$. Then the folded MI-graph $\gamma * p$ is given in Fig. 14.4. Note that $\gamma * p$ is not a recombination MI-graph, since the valency of $p$ is 6. Nevertheless, $\gamma * p$ is still balanced.                                                               $\square$

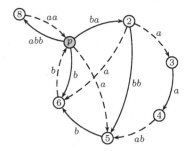

**Fig. 14.4.** The $p$-folded MI-graph $\gamma * p$ of the MI-graph $\gamma$ of Fig. 14.1. The folding is performed with respect to the $\gamma$-pointer $p = \{1, 7\}$. The new vertex $p \in V_{\gamma*p}$ is *shaded* in the figure.

If $p_1$ and $p_2$ are disjoint $\gamma$-pointers, then let $\gamma * p_1 * p_2 = (\gamma * p_1) * p_2$. More generally, for a $\gamma$-pointer family $P = \{p_1, p_2, \ldots, p_m\}$, we let

$$\gamma * p_1 * \ldots * p_m = (\gamma * p_1 * \ldots * p_{m-1}) * p_m.$$

Because the foldings on disjoint $\gamma$-pointers are "independent" of each other, the following holds.

**Lemma 14.3.** *If $p_1$ and $p_2$ are disjoint $\gamma$-pointers for an MI-graph $\gamma$, then $\gamma * p_1 * p_2 = \gamma * p_2 * p_1$.*

*Proof.* Let $p_1$ and $p_2$ be disjoint $\gamma$-pointers, and let $\gamma_1 = \gamma * p_1 * p_2$ and $\gamma_2 = \gamma * p_2 * p_1$. If an edge $e \in E_\gamma$ does not have a vertex from $p_1$ as its end, then $\varepsilon_{\gamma_1}(e) = \varepsilon_{\gamma * p_2}(e) = \varepsilon_{\gamma_2}(e)$. Similarly, if $e$ does not have an end in $p_2$, then $\varepsilon_{\gamma_1}(e) = \varepsilon_{\gamma * p_1}(e) = \varepsilon_{\gamma_2}(e)$. Suppose that $\varepsilon_\gamma(e) = (x, y)$, where $x \in p_1$ and $y \in p_2$. In this case, $\varepsilon_{\gamma * p_1}(e) = (p_1, y)$, and so $\varepsilon_{\gamma_1}(e) = (p_1, p_2)$. Similarly, $\varepsilon_{\gamma_2}(e) = (p_1, p_2)$, and so the lemma holds.    □

For a $\gamma$-pointer family $P$, we define the *$P$-**folded MI-graph*** by

$$\gamma * P = \gamma * p_1 * \ldots * p_m,$$

where $(p_1 \, p_2 \, \ldots \, p_m)$ is an arbitrary permutation of $P$. By Lemma 14.3, $\gamma * P$ is well defined. In particular, we have that

$$(\gamma * P) * P_2 = \gamma * (P_1 \cup P_2), \tag{14.5}$$

whenever $P_1$ and $P_2$ are disjoint $\gamma$-pointer families of $\gamma$.

## 14.3 Unfolding Paired MI-graphs

Let $\gamma$ be an even MI-graph together with a pairing function $\chi$ of its edges. For a vertex $x \in V_\gamma$, let $(e_{11}, e_{12}, e_{21}, e_{22}, \ldots, e_{m1}, e_{m2})$ be an ordering of the incoming edges $E_\gamma^+(x)$ such that $\chi(e_{i1}) = \bar{e}_{i2}$, and hence also $\chi(e_{i2}) = \bar{e}_{i1}$. Also, let $x^1, x^2, \ldots, x^m$ be new vertices that are associated with the old vertex $x$. Then the *$\chi$-**unfolded MI-graph*** $\gamma$ at $x$ is obtained from $\gamma$ by splitting the vertex $x$ into the parts $x^1, x^2, \ldots, x^m$ such that the edges $e_{i1}$ and $e_{i2}$ are incoming to $x^i$ (Fig. 14.5).

More formally, we have

$$\gamma \diamond_\chi x = (V, E_\gamma, \varepsilon, f_\gamma, h_\gamma),$$

with $V = (V_\gamma \setminus \{x\}) \cup \{x^1, \ldots, x^m\}$ such that for each $e \in E_\gamma$, $\varepsilon(e) = \varepsilon_\gamma(e)$, if $x$ is not an end of $e$, and

$$\varepsilon(e_{ij}) = \begin{cases} (y, x^i), & \text{if } \varepsilon_\gamma(e_{ij}) = (y, x) \text{ with } y \neq x, \\ (x^i, x^i), & \text{if } \varepsilon_\gamma(e_{ij}) = (x, x), \end{cases}$$

for $i \in [1, m]$ and $j = 1, 2$.

Notice that the $\chi$-unfolded MI-graph has the same edge set and the same labelling and coloring functions as $\gamma$. Note also that the pairing function $\chi$ remains a pairing function of $\gamma \diamond_\chi x$.

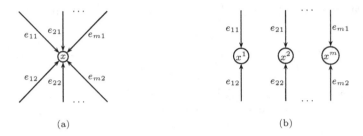

(a)                                                    (b)

**Fig. 14.5. a,b.** Unfolding with respect to a pairing function. In **a** the graph $\gamma$ is sketched with respect to the vertex $x$. The function $\chi$ pairs $e_{i1}$ and $\bar{e}_{i2}$ for each $i$. In **b** the unfolding is illustrated.

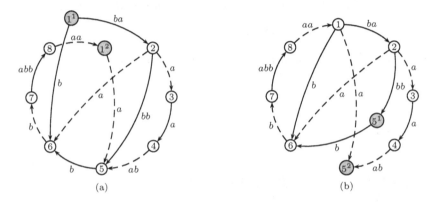

(a)                                                    (b)

**Fig. 14.6. a,b.** Unfolded MI-graphs for the MI-graph $\gamma$ in Example 14.7. In **a** the $\eta$-unfolding is performed with respect to the vertex 1, which is divided into two new vertices $1^1$ and $1^2$ that are *shaded* in the figure. In **b** the unfolding is done with respect to the vertex 5, which again is split into two parts $5^1$ and $5^2$ corresponding to the colors of the incoming edges.

*Example 14.7.* Consider again the recombination MI-graph $\gamma$ of Fig. 14.1, with the natural pairing function $\eta$. Then $\gamma \diamond_\eta 1$ is given in Fig. 14.6a, and $\gamma \diamond_\eta 5$ is given in Fig. 14.6b.                                                    □

The following result is a direct consequence of the unfolding construction.

**Lemma 14.4.** *Let $\gamma$ be an even MI-graph with a pairing function $\chi$, and let $x \in V_\gamma$. Then $\mathrm{val}_{\gamma \diamond_\chi x}(x^i) = 2$ for each new vertex $x^i$ of $\gamma \diamond_\chi x$, and $\mathrm{val}_{\gamma \diamond_\chi x}(y) = \mathrm{val}_\gamma(y)$, otherwise. Moreover, if $\gamma$ is a recombination MI-graph, then so is $\gamma \diamond_\chi x$.*

For different vertices $x_1, x_2 \in V_\gamma$, we write $\gamma \diamond_\chi x_1 \diamond_\chi x_2 = (\gamma \diamond_\chi x_1) \diamond_\chi x_2$, and, inductively, $\gamma \diamond_\chi x_1 \diamond_\chi \dots \diamond_\chi x_m = (\gamma \diamond_\chi x_1 \diamond_\chi \dots \diamond_\chi x_{m-1}) \diamond_\chi x_m$, where $x_i \neq x_j$ for all $i \neq j$.

Unfolding is commutative in the sense of the following lemma.

**Lemma 14.5.** *Let $x, y \in V_\gamma$ be different vertices of an even MI-graph $\gamma$ with a pairing function $\chi$. Then $\gamma \diamond_\chi x \diamond_\chi y = \gamma \diamond_\chi y \diamond_\chi x$.*

*Proof.* Let $x, y \in V_\gamma$ be such that $x \neq y$, and let $\gamma_1 = \gamma \diamond_\chi x \diamond_\chi y$ and $\gamma_2 = \gamma \diamond_\chi y \diamond_\chi x$. If an edge $e \in E_\gamma$ does not have $x$ as its end, then $\varepsilon_{\gamma_1}(e) = \varepsilon_{\gamma \diamond y}(e) = \varepsilon_{\gamma_2}(e)$. Similarly, if $y$ is not an end of $e$, then $\varepsilon_{\gamma_1}(e) = \varepsilon_{\gamma \diamond x}(e) = \varepsilon_{\gamma_2}(e)$. Suppose that $\varepsilon_\gamma(e) = (x, y)$. Then $\varepsilon_{\gamma_1}(e) = (x^i, y^j)$ for the vertices $x^i$ and $y^j$ determined by the pairing function $\chi$. Since $\chi$ is the same function for $\gamma \diamond x$ and $\gamma \diamond y$ (and for $\gamma$), it follows that also $\varepsilon_{\gamma_2}(e) = (x^i, y^j)$. □

By Lemma 14.5, for an even MI-graph $\gamma$ with a pairing function $\chi$ and a subset $A \subseteq V_\gamma$, we can write

$$\gamma \diamond_\chi A = \gamma \diamond_\chi x_1 \diamond_\chi x_2 \diamond_\chi \ldots \diamond_\chi x_m,$$

where $(x_1 \ldots x_m)$ is any permutation of $A$. Also, let $F(\gamma)$ be the set of all vertices of $\gamma$ with valencies at least 4:

$$F(\gamma) = \{x \in V_\gamma \mid \mathrm{val}_\gamma(x) \geq 4\}.$$

Then the MI-graph $\gamma \diamond_\chi F(\gamma)$ is called the $\chi$-**unfolded MI-graph** of $\gamma$.

**Lemma 14.6.** *If $\gamma$ is an even MI-graph with a pairing function $\chi$, then its $\chi$-unfolded MI-graph is a disjoint union of cyclic MI-graphs.*

*Proof.* For each vertex $x$ of $\gamma' = \gamma \diamond_\chi F(\gamma)$, either $\mathrm{val}_{\gamma'}(x) = 0$ or 2. This follows from Lemma 14.4 and from the fact that the valency of each $x \notin F(\gamma)$ satisfies this property. Thus the lemma follows. □

*Example 14.8.* We continue Example 14.7 for the recombination MI-graph $\gamma$ (with the natural pairing function $\eta$). The $\eta$-unfolded MI-graph of $\gamma$ is then given in Fig. 14.7. The resulting $\eta$-unfolded MI-graph consists of two disjoint cycles; one cycle consists of the four vertices $1^1$, $2^1$, $5^1$, and $6^1$, and the other one consists of the eight vertices $1^2$, $2^2$, 3, 4, $5^2$, $6^2$, 7, and 8. □

Let $\gamma$ be an MI-graph with a $\gamma$-pointer family $P$, and let $\chi$ be a pairing function of the $P$-folded MI-graph $\gamma * P$. We denote

$$\gamma \circledast_\chi P = (\gamma * P) \diamond_\chi P.$$

Thus $\gamma \circledast_\chi P$ is obtained from $\gamma$ by first folding with respect to the $\gamma$-pointer family $P$ and then unfolding with respect to the set of vertices $P$ of the folded MI-graph. We write simply $\gamma \circledast P$ for $\gamma \circledast_\eta P$ if $\gamma * P$ is a recombination MI-graph (here $\eta$ is the natural pairing function).

The following lemma states that the combined operation $\circledast_\chi P$ preserves the property of "being a disjoint union of cyclic MI-graphs."

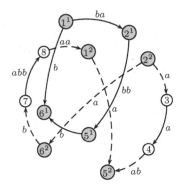

**Fig. 14.7.** The $\eta$-unfolded MI-graph of Example 14.8. It is obtained by unfolding each vertex of valency 4 of the graph $\gamma$ (Fig. 14.1). Notice that the result is a disjoint union of cyclic graphs.

**Lemma 14.7.** *Let $\gamma$ be a disjoint union of cyclic MI-graphs, $P$ be a $\gamma$-pointer family, and $\chi$ be a pairing function of the $P$-folded MI-graph $\gamma * P$. Then $\gamma \circledast_\chi P$ is a disjoint union of cyclic MI-graphs.*

*Proof.* By Lemma 14.6, $(\gamma * P) \diamond_\chi F(\gamma * P)$ is a disjoint union of cyclic MI-graphs. Since the valencies in $\gamma$ are either 0 or 2, we have $F(\gamma * P) = P$, and the lemma follows from this.    □

In the following example we describe the molecular operations *ld*, *hi*, and *dlad* by the folding and unfolding operations.

*Example 14.9.* In the figures of this example, a thick line with an arrow represents an MDS, a dashed line with an arrow represents an IES, and a light line with an arrow represents an MDS arrangement that may contain both MDSs and IESs.

(1) We consider first the molecular *ld* operation. In Fig. 14.8a we have presented an MDS arrangement $\alpha = \alpha_1 M_i M_{i+1} \alpha_2$ as a cyclic MI-graph $\gamma$ (with a "closing vertex" 0). The outgoing pointer of $M_i$ and the incoming pointer of $M_{i+1}$ are same molecular sequences, but in the figure we have denoted them "positionally" by $x$ and $y$, respectively. The $\gamma$-pointer in this case is selected to be $p = \{x, y\}$. Figure 14.8b gives the folded MI-graph $\gamma * p$ representing the loop folding of the *ld*-operation. The graph $\gamma \circledast p$ in Fig. 14.8c represents the result of recombination on pointer $p$ that splices together the MDSs $M_i$ and $M_{i+1}$. In Fig. 14.8c both of the MI-graphs are cyclic, but the rightmost MI-graph represents a linear molecule obtained by removing the auxiliary vertex 0.

(2) We now consider the *hi* operation. In Fig. 14.9a an MDS arrangement $\alpha = \alpha_1 M_i \alpha_2 \overline{M}_{i+1} \alpha_3$ is represented by a cyclic MI-graph $\gamma$. In order to simplify the figures, this time we have not indicated the auxiliary "closing vertex" 0. (Note the opposite directions of $M_i$ and $M_{i+1}$.) Again, the $\gamma$-pointer is selected

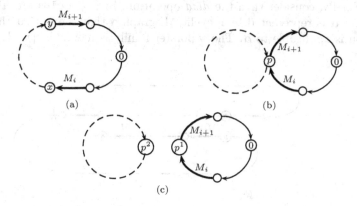

**Fig. 14.8. a–c.** The *ld* operation represented by MI-graphs. The starting situation is depicted in **a** for the MI-graph $\gamma$. In **b** the folding has been performed to obtain $\gamma * p$, and then the unfolded MI-graph $\gamma \circledast p$ is given in **c**.

to be $p = \{x, y\}$, where $x$ corresponds to the outgoing pointer of the MDS $M_i$, and $y$ corresponds to the incoming pointer of $M_{i+1}$. Then in Fig. 14.9b the graph $\gamma * p$ represents the hairpin folded molecule, and the MI-graph $\gamma \circledast p$ in Fig. 14.9c represents the recombination on pointer $p$ that splices together the MDSs $M_i$ and $M_{i+1}$. (Note that now the directions of $M_i$ and $M_{i+1}$ are the same.)

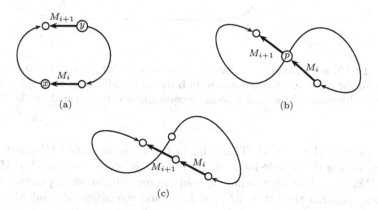

**Fig. 14.9. a–c.** The *hi* operation represented by MI-graphs. In **a** the starting situation in $\gamma$ is depicted. Note the opposite directions along $\gamma$ of the edges labelled by $M_i$ and $M_{i+1}$. In **b** the folded MI-graph $\gamma * p$ is formed, and in **c** the unfolded MI-graph $\gamma \circledast p$ is presented.

(3) Finally, consider then the *dlad* operation. In Fig. 14.10a an MDS arrangement $\alpha$ is represented by a cyclic MI-graph $\gamma$ that shows that the *dlad* operation is applicable to $\alpha$. The $\gamma$-pointer family is now $P = \{p, q\}$, where

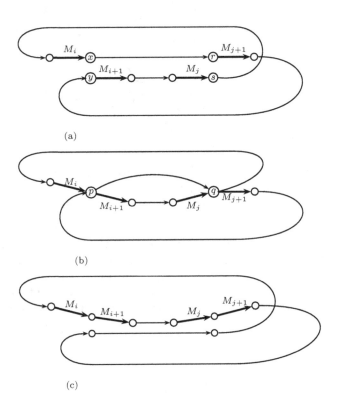

(a)

(b)

(c)

**Fig. 14.10. a–c.** The *dlad* operation represented by MI-graphs. The starting situation for the MI-graph $\gamma$ is given in **a**. In **b** the folded MI-graph $\gamma * p * q$ is made, and finally, the result of the double-loop recombination is represented by the MI-graph in **c**.

$p = \{x, y\}$ and $q = \{r, s\}$. Then Fig. 14.10b gives the folded MI-graph $\gamma * p * q$ representing the double-loop folding of the *dlad* operation, and the MI-graph $\gamma \circledast P$ in Fig. 14.10c represents the result of recombination on pointers $p, q$ that splices together the MDSs $M_i$ and $M_{i+1}$ and the MDSs $M_j$ and $M_{j+1}$.      □

## 14.4 Assembled MI-graphs of Genomes

For the rest of this chapter, we shall restrict ourselves to recombination MI-graphs. Recall that these MI-graphs have valencies either 2 or 4, and each vertex of valency 4 is balanced.

Let $\gamma$ be a cyclic MI-graph with the vertex set $V_\gamma = \{x_1, x_2, \ldots, x_n\}$ ordered accordingly around the cycle so that the edge set is

$$E_\gamma = \{e_1, \ldots, e_n, \overline{e}_1, \ldots, \overline{e}_n\}, \quad \text{where } \varepsilon_\gamma(e_i) = (x_i, x_{i+1}),$$

such that $x_{n+1} = x_1$.

A vertex $x_i$ is a **boundary vertex** of $\gamma$ if the incoming and the outgoing edges have different colors, i.e., $h_\gamma(e_{i-1}) \neq h_\gamma(e_i)$ where $e_0 = e_n$. A path $\pi$ in $\gamma$ is a **segment** if the ends of $\pi$ are boundary vertices and $h_\gamma(e) = 1$ for each edge $e$ on $\pi$. Intuitively, a segment corresponds to a (composite) MDS sequence in a gene. Notice that $h_\gamma(e) = h_\gamma(e')$ for all $e, e' \in E_\gamma$ if and only if $\gamma$ has no boundary vertices.

Let then $\gamma = \sum_{i=1}^m \gamma_i$ be a disjoint union of cyclic MI-graphs $\gamma_i$. Then each boundary vertex of each component $\gamma_i$ is said to be a **boundary vertex** of $\gamma$. Moreover, a $\gamma$-pointer family $P$ is a **boundary pointer family**, if both ends of each $p \in P$ are boundary vertices of $\gamma$.

A pair $\mathcal{G} = (\gamma, P)$ is called a **genome** if $\gamma = \sum_{i=1}^m \gamma_i$ is a disjoint union of cyclic MI-graphs $\gamma_i$ for $i \in [1, m]$, and $P$ is a boundary pointer family of $\gamma$.

*Remark 14.2.* The micronuclear DNA molecules of ciliates are linear. However, our choice to consider circular MI-graphs (corresponding to circular DNA molecules) is not a restriction, since each *linear MI-graph*, i.e., an MI-graph that is a path, can be closed to a cyclic MI-graph as follows. Let $V_\gamma = \{x_1, x_2, \ldots, x_n\}$ be ordered such that $E_\gamma = \{e_1, e_2, \ldots, e_{n-1}, \overline{e}_1, \overline{e}_2, \ldots, \overline{e}_{n-1}\}$, where $e_i = (x_i, x_{i+1})$ for $i \in [1, n-1]$. Let $\gamma'$ be the cyclic MI-graph, which is obtained from $\gamma$ by adding a vertex $x_0$ and the edges $\{e_0, e_n, \overline{e}_0, \overline{e}_n\}$ such that $\varepsilon_{\gamma'}(e_0) = (x_0, x_1)$ and $\varepsilon_{\gamma'}(e_n) = (x_n, x_0)$, $f_{\gamma'}(e_0) = \Lambda = f_{\gamma'}(e_n)$ and $h_{\gamma'}(e_0) = 2 = h_{\gamma'}(e_n)$. The new vertex $x_0$ is not a boundary vertex of $\gamma'$, and hence it is never changed by folding and unfolding operations. Therefore, in the assembled genome $x_0$ remains as a nonboundary vertex, and when it is removed from the unfolded MI-graph $\gamma \circledast P$, a linear MI-graph is recovered from the corresponding cyclic MI-graph.    □

*Example 14.10.* The pair $(\gamma, P)$ with $\gamma$ given in Fig. 14.11, and $P = \{p, q\}$, where $p = \{2, 10\}$ and $q = \{6, 9\}$, is a genome. The labels of $\gamma$ satisfy $\overline{a} = b$. Note that every vertex of $\gamma$ except 5 is a boundary vertex of $\gamma$.    □

Let $\mathcal{G} = (\gamma, P)$ be a genome, and let $R \subseteq P$. The $R$-**assembled version** of $\mathcal{G}$ is the pair

$$A(\mathcal{G}, R) = (\gamma \circledast R, P \setminus R).$$

Thus the $\gamma$-pointers in $P \setminus R$ are not used (they are "dormant") during the assembly of $A(\mathcal{G}, R)$. The **assembled genome** of $\mathcal{G}$ is the $P$-assembled version of $\mathcal{G}$ (thus all $\gamma$-pointers in $P$ are used), and it is denoted by $A(\mathcal{G})$. Two genomes $\mathcal{G}$ and $\mathcal{G}'$ are **equivalent** if they have the same assembled genome, $A(\mathcal{G}) = A(\mathcal{G}')$.

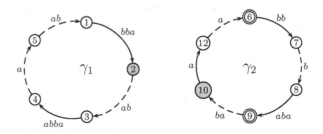

**Fig. 14.11.** The genome $(\gamma, P)$ of Example 14.10 consisting of two circular MI-graphs $\gamma_1$ and $\gamma_2$, and the boundary pointer family $P = \{p, q\}$. The vertices in $p = \{2, 10\}$ have been *shaded*, and those of $q = \{6, 9\}$ have been represented as *doubled circles*.

**Theorem 14.4.** *Let* $\mathcal{G} = (\gamma, P)$ *be a genome, and let* $R \subseteq P$. *Then* $A(\mathcal{G}, R)$ *is a genome. In particular,* $A(\mathcal{G})$ *is a genome.*

*Proof.* Since, by Lemma 14.7, $\gamma \circledast R$ is a disjoint union of cyclic MI-graphs, and hence the claim follows.                                                              □

Let $\mathcal{G} = (\gamma, P)$ be a genome. Each segment of the unfolded MI-graph $\gamma \circledast P$ is a **noncircular gene** of $\mathcal{G}$, and each cyclic component of $\gamma \circledast P$ with only edges of color 1 is a **circular gene** of $\mathcal{G}$. Hence, in general, the set of genes of $\mathcal{G}$ consists of the noncircular and the circular genes.

Each gene $g$ of a genome $\mathcal{G}$ that is not a circular gene already, gets assembled from various segments of $\mathcal{G}$, which are called the **parts** of the gene. These parts of $g$ can lie on different cycles of $\mathcal{G}$.

*Example 14.11.* Let $\mathcal{G}$ be the genome from Example 14.10 (Fig. 14.11). The $P$-folded MI-graph $\gamma * P$ is given in Fig. 14.12a. As the result of the unfolding, we obtain the MI-graph $\gamma \circledast P$ in Fig. 14.12b. All three genes $g_1, g_2, g_3$ (segments) of $\mathcal{G}$ are noncircular. They are described by the paths $g_1 \colon 1 \to p^1 \to 11$, $g_2 \colon 8 \to q^1 \to 7$, and $g_3 \colon 3 \to 4$. The labels of $g_1, g_2, g_3$ are *bbaa*, *ababb*, and *abba*, respectively.                                                              □

## 14.5 Intracyclic Unfolding

For a genome $\mathcal{G} = (\gamma, P)$, the $P$-folded MI-graph $\gamma * P$ is a recombination MI-graph that has no more connected components than $\gamma$. The MI-graph $\gamma * P$ has fewer connected components than $\gamma$ only when there exists a $\gamma$-pointer $p \in P$ whose ends lie in different connected components of $\gamma$. We will show that each recombination MI-graph $\gamma$ with $t$ connected components can be obtained from a set of $t$ cyclic MI-graphs by folding. From the genome assembly point of view this means that for each genome $\mathcal{G} = (\gamma, P)$ there

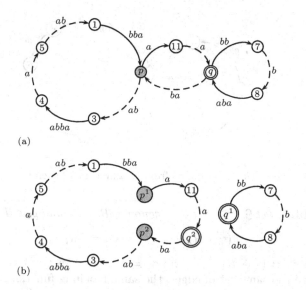

**Fig. 14.12. a** The folded MI-graph $\gamma * P$, and **b** the unfolded MI-graph $\gamma \circledast P$ of Example 14.10, where there are three segments ($g_1 : 1 \rightarrow p^1 \rightarrow 11$, $g_2 : 8 \rightarrow q^1 \rightarrow 7$, and $g_3 : 3 \rightarrow 4$) representing genes.

exists an equivalent genome $\mathcal{G}' = (\gamma', P')$ (i.e., $\gamma * P = \gamma' * P'$) such that all parts of each gene $g$ of $\mathcal{G}'$ are on one cyclic component of $\gamma'$ (because the number of connected components is equal to the number of cyclic MI-graphs).

**Theorem 14.5.** *Let $\gamma$ be a connected recombination MI-graph. There exists a cyclic MI-graph $\gamma'$ and a boundary pointer family $P$ of $\gamma'$ such that $\gamma' * P = \gamma$.*

*Proof.* By Lemma 14.2, $\gamma$ has a closed Euler trail $\pi = e_1 e_2 \ldots e_n$ that alternates on valency 4 vertices. Let $\varepsilon_\gamma(e_i) = (x_i, x_{i+1})$, where $x_1 = x_{n+1}$, and let $\chi$ be the pairing function of $\gamma$ defined by $\chi(e_i) = e_{i+1}$ for all $i \in [1, n]$ (where $e_{n+1} = e_1$). Since $\pi$ is a closed Euler trail, $\chi$ is well defined. Now, by Lemma 14.6, $\gamma \diamond_\chi F(\gamma)$ is a disjoint union of cyclic MI-graphs. By the choice of $\chi$, $\gamma \diamond_\chi F(\gamma)$ is connected, and since $\chi$ maps each edge to a differently colored edge, the new vertices $x^i$ for each $x \in F(\gamma)$ are boundary vertices of $\gamma'$. The claim follows now easily, because $(\gamma \diamond_\chi F(\gamma)) * P = \gamma$ for $P = \{\{x^1, x^2\} \mid x \in F(\gamma)\}$. □

*Example 14.12.* Let $\gamma$ be the recombination MI-graph from Fig. 14.12. Then $\gamma = \gamma' * P$ for the cyclic MI-graph $\gamma'$ in Fig. 14.13, and $P = \{\{p^1, p^2\}, \{q^1, q^2\}\}$. □

For a genome $\mathcal{G} = (\gamma, P)$, a sequence $\mathcal{S} = (P_1, P_2, \ldots, P_m)$ of subsets of $P$ is an **assembly strategy** for $\mathcal{G}$ if $\{P_1, \ldots, P_m\}$ is a partition of $P$, i.e., the $P_i$'s are mutually disjoint subsets such that $P = \cup_{i=1}^m P_i$.

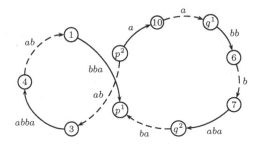

**Fig. 14.13.** The cyclic MI-graph $\gamma'$ from Example 14.12 such that $\gamma = \gamma' * P$

**Lemma 14.8.** *Let $\mathcal{G} = (\gamma, P)$ be a genome, $R \subseteq P$, and $p_0 \in P \setminus R$. Then*

$$(\gamma \circledast R) \circledast p_0 = \gamma \circledast (R \cup \{p_0\}).$$

*Proof.* Let $\gamma_1 = (\gamma \circledast R) \circledast p_0$, and $\gamma_2 = \gamma \circledast (R \cup \{p_0\})$. Both MI-graphs $\gamma_1$ and $\gamma_2$ have the same set of edges, the same labelling function, and the same coloring function as $\gamma$. Therefore it is sufficient to prove that for all $e \in E_{\gamma_1}$, $\varepsilon_{\gamma_1}(e) = \varepsilon_{\gamma_2}(e)$.

Let $R = \{p_1, p_2, \ldots, p_m\}$, where $p_i = \{x_i, y_i\}$ for $i \in [1, m]$. Thus the set of the new vertices of $\gamma_1$ and of $\gamma_2$ is $A = \{p_i^1, p_i^2 \mid i \in [1, m]\}$.

Let $\varepsilon_{\gamma_1}(e) = (x, y)$ for an edge $e \in E_{\gamma_1}$.

First of all, if $x \neq p_0^r$ and $y \neq p_0^r$ for $r = 1$ and $2$, then $\varepsilon_{\gamma \circledast R}(e) = (x, y)$, and thus also $\varepsilon_{\gamma_2}(e) = (x, y)$, because $p_0^1$ and $p_0^2$ are not ends of $e$.

Assume then that $\varepsilon_{\gamma_1}(e) = (p_i^r, p_0^r)$ for some $1 \leq i \leq m$ and $r \in \{1, 2\}$, and so $h_\gamma(e) = r$. Now, $\varepsilon_{(\gamma \circledast R) * p_0}(e) = (p_i^r, p_0)$, $\varepsilon_{\gamma \circledast R}(e) = (p_i^r, z_0)$, where $z_0 \in \{x_0, y_0\}$, and hence $\varepsilon_{\gamma * R}(e) = (p_i, z_0)$. Consequently, $\varepsilon_{(\gamma * R) * p_0}(e) = (p_i, p_0)$, where $(\gamma * R) * p_0 = \gamma * (R \cup \{p_0\})$ by (14.5). Therefore, $\varepsilon_{\gamma \circledast (R \cup \{p_0\})}(e) = (p_i^r, p_0^r)$, since $h_{\gamma * (R \cup \{p_0\})}(e) = r$. Hence, also in this case, $\varepsilon_{\gamma_1}(e) = \varepsilon_{\gamma_2}(e)$.

The case where $\varepsilon_{\gamma_1}(e) = (x, p_0^r)$ for some $x \in V_\gamma$ is similar, but somewhat easier than the above case. Finally note that if $r \neq s$, there are no edges having the ends $p_i^r$ and $p_j^s$ for any $i$ and $j$. Hence the lemma holds. □

Lemma 14.8 gives inductively the following generalization.

**Lemma 14.9.** *Let $\mathcal{G} = (\gamma, P)$ be a genome, and let $P_1, P_2 \subseteq P$ be disjoint. Then $(\gamma \circledast P_1) \circledast P_2 = \gamma \circledast (P_1 \cup P_2) = (\gamma \circledast P_2) \circledast P_1$.*

For disjoint boundary pointer families $P_1, P_2, \ldots, P_m \subseteq P$ of an MI-graph $\gamma$, Lemma 14.9 allows us to write $\gamma \circledast P_1 \circledast P_2 \circledast \ldots \circledast P_m = (\ldots ((\gamma \circledast P_1) \circledast P_2) \circledast \ldots) \circledast P_m$. In particular,

$$\gamma \circledast P_1 \circledast P_2 \circledast \ldots \circledast P_m = \gamma \circledast \bigcup_{i=1}^{m} P_i. \qquad (14.6)$$

Lemma 14.9 together with condition (14.6) guarantees that every assembly strategy produces the same assembled genome:

**Theorem 14.6.** *For every assembly strategy* $\mathcal{S} = (P_1, P_2, \ldots, P_m)$ *for a genome* $\mathcal{G} = (\gamma, P)$, *we have* $\gamma \circledast P = \gamma \circledast P_1 \circledast P_2 \circledast \ldots \circledast P_m$.

In particular, for any genome $\mathcal{G} = (\gamma, P)$ with $P = \{p_1, p_2, \ldots, p_m\}$, the assembly strategy $\mathcal{S} = (\{p_1\}, \{p_2\}, \ldots, \{p_m\})$ consisting of the singleton sets yields the assembled genome $A(\mathcal{G})$: $\gamma \circledast p_1 \circledast \ldots \circledast p_m = \gamma \circledast P$. However, such an assembly strategy can be "intercircular" in the sense that parts of a gene $g$ of $\mathcal{G}$ that lie in the same connected component of $\gamma \circledast \{p_1, \ldots, p_i\}$ can be in different connected components of $\gamma \circledast \{p_1, \ldots, p_{i+1}\}$ (and they are "reunited" in the same connected component in the final MI-graph $\gamma \circledast P$). We shall look now for simple assembly strategies for genomes that are "intracyclic", i.e., the above does not happen.

A boundary pointer family $R \subseteq P$ of a genome $\mathcal{G} = (\gamma, P)$ is called **intracyclic** if for every gene $g$ of $\mathcal{G}$, any two parts $g'$ and $g''$ of $g$ that lie in the same connected component of $\gamma$ lie in the same connected component of $\gamma \circledast R$.

Furthermore, we say that two boundary vertices $x, y$ of a cyclic MI-graph $\gamma$ are **opposite** if $\gamma = \pi_1 \pi_2 \pi_3 \pi_4 \pi_5$, where $\pi_2$ and $\pi_4$ are different segments of $\gamma$ such that $\iota(\pi_2) = x$ and $\tau(\pi_4) = y$, or $\tau(\pi_2) = x$ and $\iota(\pi_4) = y$. If $x$ and $y$ are not opposite, they are **similar**. Note that being similar includes the case where the two vertices are the ends of one segment.

We begin by considering intracyclicity for singleton pointer families. First, we consider the case when the ends of the pointer are in different connected components of the genome, or they are similar and in the same connected component.

**Lemma 14.10.** *Let* $\mathcal{G} = (\gamma, P)$ *be a genome, and let* $p \in P$. *If the ends of* $p$ *are either*

*(i) in different connected component of* $\gamma$, *or*
*(ii) similar in the same connected component of* $\gamma$,

*then* $\{p\}$ *is intracyclic.*

*Proof.* Let $p = \{x, y\}$, where $x$ and $y$ are different boundary vertices.

Case (i) is illustrated in Fig. 14.14. In Fig. 14.14a the connected components $\gamma_1$ and $\gamma_2$ of $\gamma$ containing $x$ and $y$, respectively, are shown together with the segments for which $x$ and $y$ are the initial vertices. Then b gives $(\gamma_1 + \gamma_2) * p$, and c shows the connected component $(\gamma_1 + \gamma_2) \circledast p$ of $\gamma \circledast p$. Since none of the connected components of $\gamma$ is disconnected into two or more circular MI-graphs during the process, it follows that $\{p\}$ is intracyclic.

In Case (ii) we consider two subcases, (ii.1) and (ii.2), given by Figs. 14.15 and 14.16, respectively.

In Case (ii.1), $x$ and $y$ are vertices of different segments as illustrated in Fig. 14.15a. Then folding of $\gamma_1$ on $p$ yields $\gamma_1 * p$, see Fig. 14.15b, and the unfolding on $p$ yields $\gamma_1 \circledast p$, see Fig. 14.15c. In this way, no connected component of $\gamma$ is disconnected during the process.

In Case (ii.2), $x$ and $y$ are the two boundary vertices of the same segment as illustrated in Fig. 14.16a. The folding of $\gamma_1$ on $p$ yields $\gamma_1 * p$ given in Fig. 14.16b, and then the unfolding on $p$ yields $\gamma_1 \circledast p$ given in Fig. 14.16c. If the segment delimited by $x$ and $y$ in $\gamma_1$ is a part of a gene $g$ of the genome, then this part must be the whole $g$, because it forms a connected component of $\gamma_1 \circledast p$ that has no boundary vertices (and hence will not be combined any more with any other connected components). The connected components of $\gamma$ other than $\gamma_1$ remain as they were. The lemma follows from the conclusions of Cases (i) and (ii).                                   □

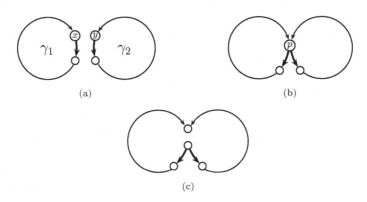

(a)                                     (b)

(c)

**Fig. 14.14. a–c.** One pointer cases of Lemma 14.10(i). In **a** the ends of the $\gamma$-pointer $p = \{x, y\}$ are in different connected components $\gamma_1$ and $\gamma_2$ of the MI-graph $\gamma$. In **b** we have the folded MI-graph $(\gamma_1 + \gamma_2) * p$, and in **c** the unfolded MI-graph $(\gamma_1 + \gamma_2) \circledast p$.

**Lemma 14.11.** *Let $\mathcal{G} = (\gamma, P)$ be a genome, and $p = \{x, y\} \in P$ be a $\gamma$-pointer such that $x$ and $y$ are opposite and on the same cyclic MI-graph $\gamma_i$ of $\gamma$. Then either $\{p\}$ is intracyclic or there exists a $\gamma$-pointer $q$ such that $\{p, q\}$ is intracyclic.*

*Proof.* In transforming $\gamma$ to $\gamma \circledast p$, the cyclic MI-subgraph $\gamma_i$ that contains $x$ and $y$ is disconnected into two cyclic MI-graphs $\gamma_i'$ and $\gamma_i''$ of $\gamma \circledast p$, see Fig. 14.17. If $\{p\}$ is not intracyclic for $\mathcal{G}$, then each of $\gamma_i'$ and $\gamma_i''$ contains a part of a gene $g$ of $\mathcal{G}$, and since $g$ is a path in $\gamma \circledast P$, there exists a $\gamma$-pointer $q = \{x', y'\}$ such that $x'$ is on $\gamma_i'$ and $y'$ is on $\gamma_i''$. By Lemma 14.10, $\{q\}$ is intracyclic for $\gamma \circledast p$, and therefore, by Lemma 14.9, $\{p, q\}$ is intracyclic for $\gamma$.                                   □

*Example 14.13.* Consider the genome $\mathcal{G} = (\gamma, P)$ from Fig. 14.18, where each of the $\gamma$-pointers $p = \{1, 6\}$ and $q = \{3, 8\}$ consists of opposite vertices. Note that $\gamma \circledast \{p, q\}$ is connected, although, as can be shown, both $\gamma \circledast p$ and $\gamma \circledast q$ are disconnected.                                   □

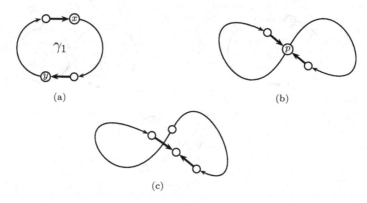

**Fig. 14.15.** One pointer case for Lemma 14.10(ii). In (a) the ends of the $\gamma$-pointer $p = \{x, y\}$ are in the same connected component $\gamma_1$ of $\gamma$, they are similar and in different segments. In (b) we have the folded MI-graph $\gamma_1 * p$, and in (c) the unfolded MI-graph $\gamma_1 \circledast p$.

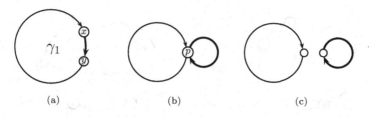

**Fig. 14.16. a–c.** One pointer case for Lemma 14.10(ii). In **a** the ends of the $\gamma$-pointer $p = \{x, y\}$ are in the same connected component $\gamma_1$ of $\gamma$ they are similar and in the same segment. In **b** we have the folded MI-graph $\gamma_1 * p$, and in **c** the unfolded MI-graph $\gamma_1 \circledast p$.

As a corollary to Lemmata 14.10 and 14.11, we have the following result.

**Theorem 14.7.** *For each genome $\mathcal{G}$, there exists an intracyclic assembly strategy $\mathcal{S} = (P_1, P_2, \ldots, P_m)$ such that $1 \leq \mathrm{card}(P_i) \leq 2$ for all $i$.*

Combining Theorem 14.7 and Theorem 14.5, one obtains for each genome that yields a connected assembled genome, an equivalent connected genome, for which there exists an intracyclic assembly strategy using no more than two pointers in each assembly step.

**Theorem 14.8.** *Let $\mathcal{G} = (\gamma, P)$ be a genome such that $\gamma \circledast P$ is connected. Then there exist an equivalent genome $\mathcal{G}' = (\gamma', P')$ and an assembly strategy $\mathcal{S} = (P_1, P_2, \ldots, P_m)$ of $\mathcal{G}'$ such that $\gamma'$ is connected, $1 \leq \mathrm{card}(P_i) \leq 2$ for all $i$, and each $\gamma \circledast \cup_{i=1}^{j} P_i$ is a cyclic MI-graph for $j \in [1, m]$.*

Our final example illustrates a real-life case of gene assembly in ciliates.

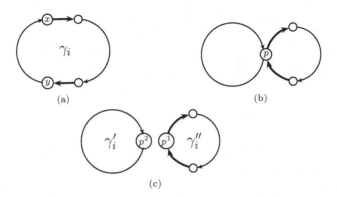

**Fig. 14.17. a–c.** One pointer case for Lemma 14.11. In **a** the ends of the $\gamma$-pointer $p = \{p, q\}$ are opposite, and in the same connected component $\gamma_i$ of $\gamma$. In **b** we have the folded MI-graph $\gamma_i * p$, and in **c** the unfolded MI-graph that is a disjoint union of two cyclic MI-graphs $\gamma_i'$ and $\gamma_i''$.

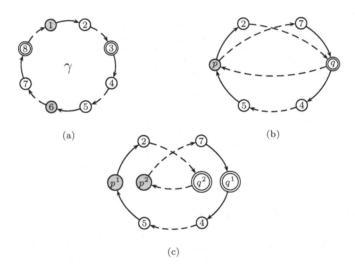

**Fig. 14.18.** In **a** we have the genome $\mathcal{G} = (\gamma, P)$ from Example 14.13 where $P = \{p, q\}$ with $p = \{1, 6\}$ (*shaded*) and $q = \{3, 8\}$ (*double circled*). In **b** we have the folded MI-graph $\gamma * P$, and in **c** the unfolded MI-graph $\gamma \circledast P$.

*Example 14.14.* Recall that the MDS/IES structure of the actin gene in *Sterkiella nova*, is described by

$$M_3 I_1 M_4 I_2 M_6 I_3 M_5 I_4 M_7 I_5 M_9 I_6 \overline{M}_2 I_7 M_1 I_8 M_8. \tag{14.7}$$

We shall consider now the assembly of this gene; its macronuclear version is given by the orthodox order of the MDSs $g = M_1 M_2 M_3 M_4 M_5 M_6 M_7 M_8 M_9$.

To present (14.7) as an MI-graph, we introduce a new vertex 0 to ensure that the MI-graph is cyclic (see Remark 14.2). The so-obtained cyclic MI-graph $\gamma$ is given in Fig. 14.19. Note that we have drawn the edge $(14, 13)$ with the label $M_2$ instead of $(13, 14)$ (with the label $\overline{M}_2$).

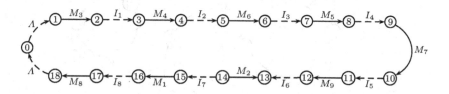

**Fig. 14.19.** A presentation of the actin gene in *Sterkiella nova* as an MI-graph. The "closing vertex" 0 is added together with the edges to 1 and from 18 (both with *label $\Lambda$*) to make the MI-graph cyclic.

The boundary pointer family $P$ of $\gamma$ consists of the following pointers:

$$p_1 = \{1, 13\}, \quad p_2 = \{2, 3\}, \quad p_3 = \{4, 7\}, \quad p_4 = \{5, 8\}, \quad p_5 = \{6, 9\},$$
$$p_6 = \{10, 17\}, \quad p_7 = \{11, 18\}, \quad \text{and} \quad p_8 = \{14, 16\}.$$

The $P$-folded MI-graph $\gamma * P$ is given in Fig. 14.20, and the final genome $\gamma \circledast P$ is given in Fig. 14.21. The segment labelled by $M_1 M_2 \ldots M_9$ is the gene of this genome. Removing the vertex 0 (together with its two adjacent edges) yields the linear structure of this component given in Fig. 14.22.

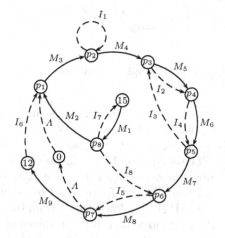

**Fig. 14.20.** The $P$-folded MI-graph of actin gene in *Sterkiella nova*

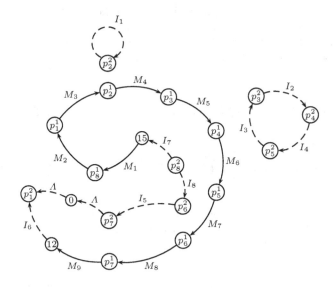

**Fig. 14.21.** The final genome of actin gene in *Sterkiella nova*

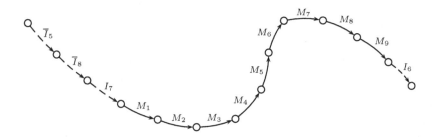

**Fig. 14.22.** The component MI-graph of actin gene

Hence, during the assembly of the actin gene in *Sterkiella nova*, the IES $I_1$ ("polluted" by a pointer) has been excised as a circular molecule; the combined IES $I_2 I_4 I_3$ (polluted by three pointers) has also been excised as a circular molecule. (These circular IESs will be digested by the host cell.) The molecule containing the assembled actin gene of *Sterkiella nova* will also contain the combined linear IES $\bar{I}_5 \bar{I}_8 I_7$ (polluted by pointers) upstream from the gene, and the (polluted) IES $I_6$ downstream from the gene. As a matter of fact, the physical actin gene molecule will be produced by an involved process that cuts off the molecule at the vertices 12 and 15, called **markers**, and attaches telomeres at these markers. □

# Notes on References

- This chapter is based on the results by Ehrenfeucht et al. [18], where, however, a more general approach was adopted. In [18] the graphs were not assumed to be bicolored that is, the edge colors were taken from an arbitrarily large set $[1, k]$.
- Theorem 14.2 is a special case of a general characterization of alternating Eulerian graphs for edge-colored graphs. This result is from Kotzig [36]. For a proof of the general result, we also refer to Pevzner [43].
- For the MI-graphs, Pevzner [42] proved that any two alternating closed Euler trails can be obtained from each other by using relatively simple transformations of closed walks. The **exchange operation** transforms a closed walk with a decomposition $\pi = \pi_1\pi_2\pi_3\pi_4\pi_5$, where $\iota(\pi_2) = \iota(\pi_4)$ and $\tau(\pi_2) = \tau(\pi_4)$, to the closed walk $\pi' = \pi_1\pi_4\pi_3\pi_2\pi_5$. The **reflection operation** transforms a closed walk with a decomposition $\pi = \pi_1\pi_2\pi_3$, where $\pi_2$ is a closed walk, to the closed walk $\pi' = \pi_1\overline{\pi}_2\pi_3$. It was proved by Pevzner that if $\gamma$ is an MI-graph, then every two alternating closed Euler trails can be transformed to each other by a finite number of exchange and reflection operations that preserve alternating closed walks.
- For graph theoretic notions concerning Eulerian graphs, including Euler's theorem, we refer to West [58].
- Theorem 14.3 is from Tucker [56], see also West [58].

# Part IV

Epilogue

# 15

# Intermolecular Model

In this chapter, we consider the *intermolecular* model for gene assembly in ciliates proposed by Landweber and Kari. However, we present this model in terms of pointer reductions, similarly as done in the rest of the book for the *intramolecular* model. We consider unary string rewriting rules that correspond to the string negative rule snr and its reverse rule, as well as a binary string rule. This model does not cover inversions, i.e., the string positive rule spr is not covered in this framework. However, we show how inversions can in fact be handled, provided that the initial string is available in multiple copies.

## 15.1 String Rules

Let $\Sigma$ be an alphabet, and let $w \in \Sigma^{\circledast}$ be a signed string. Recall that $[w]$ denotes the circular string corresponding to $w$. Since in this chapter we are concerned with multiplicity of strings, we shall use $A + B$ to denote the union of the sets $A, B$ of strings. Also, we shall often identify a singleton set $\{w\}$ with its unique element $w$. Therefore we have expressions of the form $w + u + [v]$, where $w$ and $u$ are strings, and $[v]$ is a circular string.

Consider the following unary string rewriting rule:

$$uxwxv \longrightarrow uxv + [wx], \qquad (15.1)$$

where $u, x, w, v \in \Sigma^{\circledast}$ with $x \neq \Lambda$. Thus, rule (15.1) splits a linear string into a linear string and a circular string, as illustrated in Fig. 15.1. This rule models the intramolecular recombination that occurs when a micronuclear gene undergoes assemblage to its macronuclear version. Here the direct repeat $x$ guides the homologous recombination so that the molecule undergoes a strand exchange in $x$.

We shall also consider the following string rewriting rule:

$$uxv + [wx] \longrightarrow uxwxv, \qquad (15.2)$$

which merges two strings: a linear string $uxv$ and a circular string $[wx]$. This rule reverses the effect of rule (15.1).

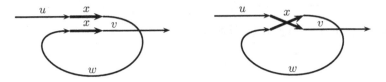

**Fig. 15.1.** Illustration of rule (15.1) for strings, where the common substring $x$ is presented as a *thick line*. By this rule a string $uxwxv$ is split to two parts: a linear string $uxv$ and a circular string $[wx]$.

Moreover, we shall consider the following binary string rewriting rule:

$$uxv + u'xv' \longrightarrow uxv' + u'xv, \tag{15.3}$$

where $u, u', v, v', x \in \Sigma^{\circledast}$, with $x \neq \Lambda$. This rule involves two linear strings, and it produces two linear strings, as illustrated in Fig. 15.2. This operation models

**Fig. 15.2.** Illustration of rule (15.3). Two strings $uxv$ and $u'xv'$ with a common substring $x$ are recombined to obtain two strings $uxv'$ and $u'xv$.

intermolecular recombination when the direct repeat $x$ guides the homologous recombination of two molecules, so that molecules undergo a strand exchange in $x$.

## 15.2 The Intermolecular Model in Terms of Signed Strings

We shall now investigate the above string rewriting rules in the formal framework of signed strings of pointers. Note that rule (15.1) generalizes in many respects the *ld*-operation for strings. Recall, however, that in the *ld*-rule between the two repeated sequences $x$ (pointers) there is only one IES.

The rules (15.1), (15.2), and (15.3) become the following reduction rules for signed strings:

$$upwpv \xrightarrow{p} uv + [w], \tag{15.4}$$

$$upv + [wp] \xrightarrow{p} uwv, \tag{15.5}$$

$$upv + u'pv' \xrightarrow{p} uv' + u'v, \tag{15.6}$$

where $u, w, v, u'$, and $v'$ are strings consisting of pointers, and $p$ is a pointer. If $w = \Lambda$, then we omit the circular empty string $[\Lambda]$ from (15.4).

Note that unlike in rules (15.1), (15.2), and (15.3), rules (15.4), (15.5), and (15.6) remove pointers involved in the applications of these rules. Therefore the rules (15.4), (15.5), and (15.6) are *irreversible rules*.

The snr-rule is clearly a special case of the rule (15.4) obtained by setting $w = \Lambda$. We shall now show how the intramolecular rules (15.4) and the intermolecular rules (15.5) can be used to simulate the string double rule sdr on signed strings.

Let $w = w_1 p w_2 q w_3 p w_4 q w_5$. Then

$$w \xrightarrow{p} w_1 w_4 q w_5 + [w_2 q w_3] = w_1 w_4 q w_5 + [w_3 w_2 q]$$
$$\xrightarrow{q} w_1 w_4 w_3 w_2 w_5 = \mathsf{sdr}_{p,q}(w).$$

The string positive rule spr is more difficult to simulate using the rules from the intermolecular model, as they do not deal with inverted pointers. Nevertheless, we shall show now that spr can in fact be expressed in the intermolecular model, *provided that the input string is available in two copies*. Moreover, *we consider all linear strings modulo inversion*. The first assumption is essentially used in research on the intermolecular model (see the notes on references at the end of this chapter). The second assumption is quite natural whenever we model double-stranded DNA molecules. As a matter of fact, we shall use the two assumptions to conclude that for each input string, both the string and its inversion are available. Then the string positive rule spr can be simulated as follows.

Let $w = w_1 p w_2 p w_3$ be given, and therefore also $\overline{w} = \overline{w}_3 p \overline{w}_2 \overline{p} \,\overline{w}_1$ is available. Then, using rule (15.6) twice, we obtain

$$w + \overline{w} \xrightarrow{p} w_1 \overline{w}_2 \,\overline{p}\,\overline{w}_1 + \overline{w}_3 \, w_2 \,\overline{p}\, w_3$$
$$\xrightarrow{\overline{p}} w_1 \overline{w}_2 w_3 + \overline{w}_3 w_2 \overline{w}_1 = \mathsf{spr}_p(w) + \overline{\mathsf{spr}_p(w)}.$$

Note that, since two copies of the initial string are available, this rule yields two copies of $\mathsf{spr}_p(w)$.

The following universality result is then a simple consequence of Theorem 9.1.

**Theorem 15.1.** *Each legal string has a successful string reduction in the intermolecular model, provided that the string is available in two copies.*

*Example 15.1.* Consider the legal string $u$ for the actin gene in *Sterkiella nova*, see Example 8.4:

$$u = 3\,4\,4\,5\,6\,7\,5\,6\,7\,8\,9\,\overline{3}\,\overline{2}\,2\,8\,9.$$

Then $u$ can be assembled in the intermolecular model as follows:

$$u \xrightarrow{4} 3\,5\,6\,7\,5\,6\,7\,8\,9\,\bar{3}\,\bar{2}\,2\,8\,9$$

$$\xrightarrow{7} 3\,5\,6\,8\,9\,\bar{3}\,\bar{2}\,2\,8\,9 + [5\,6]$$

$$\xrightarrow{8} 3\,5\,6\,9 + [9\,\bar{3}\,\bar{2}\,2] + [5\,6] = 3\,5\,6\,9 + [9\,\bar{3}\,\bar{2}\,2] + [6\,5]$$

$$\xrightarrow{5} 3\,6\,6\,9 + [9\,\bar{3}\,\bar{2}\,2]$$

$$\xrightarrow{6} 3\,9 + [9\,\bar{3}\,\bar{2}\,2] = 3\,9 + [\bar{3}\,\bar{2}\,2\,9]$$

$$\xrightarrow{9} 3\,\bar{3}\,\bar{2}\,2 \,.$$

Since also $\bar{u}$ is available, we continue as follows:

$$u + \bar{u} \to \ldots \to 3\,\bar{3}\,\bar{2}\,2 + \bar{2}\,2\,3\,\bar{3}$$

$$\xrightarrow{3} \bar{3} + \bar{2}\,2\,3\,\bar{3}\,2 \xrightarrow{\bar{3}} \bar{2}\,2 + \bar{2}\,2$$

$$\xrightarrow{2} \bar{2} + \bar{2} \xrightarrow{\bar{2}} \Lambda + \Lambda \,.$$

$\square$

## 15.3 Invariants of the Intermolecular Model

We shall consider in this section more examples of assembly in the intermolecular model. However, to compare the results of such assemblies with the results predicted by the intramolecular model, we formalize the intermolecular model in terms of MDS–IES descriptors, in the style of Sect. 12.1. Thus, in these terms, the intermolecular model will consist of these rules:

$$\delta_1(q,p)\delta_2(p,r)\delta_3 \xrightarrow{p} \delta_1(q,r)\delta_3 + [\delta_2],$$

$$\delta_1(p,q)\delta_2(r,p)\delta_3 \xrightarrow{p} \delta_1\delta_3 + [\delta_2(r,q)],$$

$$\delta_1(p,q)\delta_2 + [(r,p)\delta_3] \xrightarrow{p} \delta_1\delta_3(r,q)\delta_2,$$

$$\delta_1(q,p)\delta_2 + [(p,r)\delta_3] \xrightarrow{p} \delta_1(q,r)\delta_3\delta_2,$$

$$\delta_1(p,q)\delta_2 + \delta_3(r,p)\delta_4 \xrightarrow{p} \delta_1\delta_4 + \delta_3(r,q)\delta_2,$$

where $\delta_1, \delta_2, \delta_3 \in (\Gamma_k \cup \Omega)^*$, and $\Omega$ is the set of IESs (see Sect. 12.1).

*Example 15.2.* Consider the actin gene in *Sterkiella nova* from Example 12.4 having the MDS–IES descriptor

$$\delta = (3,4)I_1(4,5)I_2(6,7)I_3(5,6)I_4(7,8)I_5(9,e)I_6(\bar{3},\bar{2})I_7(b,2)I_8(8,9)\,.$$

Then $\delta$ can be assembled in the intermolecular model as follows:

$$\delta \xrightarrow{5} \quad (3,4)I_1(4,6)I_4(7,8)I_5(9,e)I_6(\overline{3},\overline{2})I_7(b,2)I_8(8,9) + [I_2(6,7)I_3]$$

$$\xrightarrow{8} \quad (3,4)I_1(4,6)I_4(7,9) + [I_5(9,e)I_6(\overline{3},\overline{2})I_7(b,2)I_8] + [I_2(6,7)I_3]$$

$$\xrightarrow{4} \quad (3,6)I_4(7,9) + [I_1] + [I_5(9,e)I_6(\overline{3},\overline{2})I_7(b,2)I_8] + [I_2(6,7)I_3]$$

$$\xrightarrow{7} \quad (3,6)I_4I_3I_2(6,9) + [I_1] + [I_5(9,e)I_6(\overline{3},\overline{2})I_7(b,2)I_8]$$

$$\xrightarrow{6} \quad (3,9) + [I_4I_3I_2] + [I_1] + [I_5(9,e)I_6(\overline{3},\overline{2})I_7(b,2)I_8]$$

$$\xrightarrow{9} \quad (3,e)I_6(\overline{3},\overline{2})I_7(b,2)I_8I_5 + [I_4I_3I_2] + [I_1].$$

Since $\overline{\delta}$ is also available, the assembly continues as follows. Note that for a (circular) string $\tau$, we use $2 \cdot \tau$ to denote $\tau + \overline{\tau}$:

$$\delta + \overline{\delta} \; \rightarrow \ldots \rightarrow \; (3,e)I_6(\overline{3},\overline{2})I_7(b,2)I_8I_5 + \overline{I}_5\overline{I}_8(\overline{2},\overline{b})\overline{I}_7(2,3)\overline{I}_6(\overline{e},\overline{3})$$
$$+ 2 \cdot [I_4I_3I_2] + 2 \cdot [I_1]$$

$$\xrightarrow{\overline{2}} \quad (3,e)I_6(\overline{3},\overline{b})\overline{I}_7(2,3)\overline{I}_6(\overline{e},\overline{3}) + \overline{I}_5\overline{I}_8I_7(b,2)I_8I_5$$
$$+ 2 \cdot [I_4I_3I_2] + 2 \cdot [I_1]$$

$$\xrightarrow{2} \quad (3,e)I_6(\overline{3},\overline{b})\overline{I}_7I_8I_5 + \overline{I}_5\overline{I}_8I_7(b,3)\overline{I}_6(\overline{e},\overline{3}) + 2 \cdot [I_4I_3I_2] + 2 \cdot [I_1]$$

$$\xrightarrow{\overline{3}} \quad (3,e)I_6 + \overline{I}_5\overline{I}_8I_7(b,3)\overline{I}_6(\overline{e},\overline{b})\overline{I}_7I_8I_5 + 2 \cdot [I_4I_3I_2] + 2 \cdot [I_1]$$

$$\xrightarrow{3} \quad \overline{I}_6(\overline{e},\overline{b})\overline{I}_7I_8I_5 + \overline{I}_5\overline{I}_8I_7(b,e) + 2 \cdot [I_4I_3I_2] + 2 \cdot [I_1]$$

$$= \quad 2 \cdot (\overline{I}_6(\overline{e},\overline{b})\overline{I}_7I_8I_5 + +[I_1] + [I_3I_2I_4]).$$

Note that this intermolecular assembly predicts the same context for the assembled string, the same set of residual strings, and the same linearity of the assembled string as the intramolecular assembly considered in Example 12.4.     □

*Example 15.3.* Consider the MDS–IES descriptor

$$\delta = (\overline{10},\overline{8})I_1(\overline{3},\overline{b})I_2(\overline{5},\overline{3})I_3(10,11)I_4(5,8)I_5(11,e)$$

from Examples 12.2 and 12.5. Then $\delta$ can be assembled in the intermolecular model as follows:

$$\delta \xrightarrow{3} \quad (\overline{10},\overline{8})I_1I_3(10,11)I_4(5,8)I_5(11,e) + [I_2(\overline{5},\overline{b})]$$

$$\xrightarrow{11} \quad (\overline{10},\overline{8})I_1I_3(10,e) + [I_4(5,8)I_5] + [I_2(\overline{5},\overline{b})].$$

Since also $\overline{\delta}$ is available, the assembly continues as follows:

$$\delta + \bar{\delta} \;\rightarrow\; \ldots \;\rightarrow\; (\overline{10},\overline{8})I_1I_3(10,e) + (\overline{e},\overline{10})\overline{I}_3\overline{I}_1(8,10)$$

$$+\, 2 \cdot [I_4(5,8)I_5] + 2 \cdot [I_2(\overline{5},\overline{b})]$$

$$\xrightarrow{\;\overline{10}\;}\; \overline{I}_3\overline{I}_1(8,10) + (\overline{e},\overline{8})I_1I_3(10,e) + 2 \cdot [I_4(5,8)I_5] + 2 \cdot [I_2(\overline{5},\overline{b})]$$

$$\xrightarrow{\;10\;}\; \overline{I}_3\overline{I}_1(8,e) + (\overline{e},\overline{8})I_1I_3 + [I_4(5,8)I_5] + [\overline{I}_5(\overline{8},\overline{5})\overline{I}_4] + 2 \cdot [I_2(\overline{5},\overline{b})]$$

$$\xrightarrow{\;8,\,\overline{8}\;}\; \overline{I}_3\overline{I}_1I_5I_4(5,e) + (\overline{e},\overline{5})\overline{I}_4\overline{I}_5I_1I_3 + [(\overline{5},\overline{b})I_2] + [\overline{I}_2(b,5)]$$

$$\xrightarrow{\;5,\,\overline{5}\;}\; \overline{I}_3\overline{I}_1I_5I_4\overline{I}_2(b,e) + (\overline{e},\overline{b})I_2\overline{I}_4\overline{I}_5I_1I_3$$

$$=\; 2 \cdot \overline{I}_3\overline{I}_1I_5I_4\overline{I}_2(b,e).$$

Note that, again, we obtain the same context for the assembled string, the same set of residual strings, and the same linearity of the assembled string as the intramolecular assembly considered in Example 12.5.    □

## Notes on References

- The intermolecular model originates in paper [37]. It was followed by a series of papers by L.Kari and Landweber and their collaborators including Daley, Ibarra, J.Kari; see [8, 9, 10, 32, 33, 34]. As a matter of fact, the computational nature of the gene assembly process was brought to the attention of the DNA computing community by L.Kari and Landweber in [37, 38]. The major difference between the Kari and Landweber model and the model considered in this book is that their model is intermolecular and it tries to capture both the process of identifying pointers and the process of using pointers by operations that accomplish gene assembly. In our model we assume that the pointer structure of a molecule is known, i.e., the pointers have been already identified. This implies some important differences between the models: e.g., our (legal) strings contain two occurrences for each pointer still present, and moreover processing a pointer implies its removal from the processed string; these properties do not hold in the Kari–Landweber model. Finally, the bulk of the work by Kari and Landweber is concerned with the computational power of the operations in the sense of computability theory; e.g., they prove in [34, 37, 38] that their model has the computational power of the Turing machine. On the other hand, research presented in this book deals with representations and properties of the gene assembly process (represented by various kinds of reduction systems). We believe that the two approaches together shed light on the computational nature of gene assembly in ciliates.
- The string rewriting rules (15.1) and (15.2) are from Landweber and Kari [37, 38]. Also, rule (15.3) is stated in [37, 38], however, this rule is a general *splicing operation* for strings introduced by Head [28] in 1987.
- Rules (15.1), (15.2), and (15.3) are translated into the framework of pointer reduction system through the rules (15.4), (15.5), and (15.6). Through this

translation we leave the framework of Kari and Landweber and focus on questions considered in this book: the universality of our rules, realizability, and invariants of the rules.

- The realizability of the string positive and string double rules, spr and sdr, by the reduction rules (15.4), (15.5), and (15.6) is considered for the first time in this chapter.
- The assumption of availability of the initial string in at least two copies is essential in the study of the Kari and Landweber model (see [32, 33, 34, 38]). We carried this assumption over into our considerations.
- Note that the cyclic graph decomposition model of Chap. 14 accommodates both intramolecular and intermolecular operations. This approach also explains why the intermolecular model predicts the same invariants as the intramolecular model.

# 16

# Discussion

In this book we have presented a systematic formal study of gene assembly in ciliates. After providing biological preliminaries we have formulated and studied gene assembly on different levels of abstraction. Finally, we presented in more detail three specific research areas: invariants, patterns, and the representation of gene assembly through loop decomposition of graphs. In this final chapter, we make some concluding remarks, and we point out some possible research problems.

## 16.1 Between Biology and Computer Science

The research on the computational nature of gene assembly is interdisciplinary, i.e., it is carried out by both biologists and computer scientists. It is our conviction that both disciplines have already benefitted from this research.

On the biology side, the operational model based on the three operations *ld*, *hi*, and *dlad* and their formalizations provides a convenient framework for reasoning about gene assembly. This formal framework can yield new insight into the structure of genes formed during the process of gene assembly (see Chap. 13), or novel ideas for specifying the complexity of genes in stichotrichs (see Sect. 13.4), or computer software that allows simulation/generation of possible strategies for gene assembly.

On the computer science side, the study of gene assembly has led to many novel and interesting questions and models of computational processes. More specifically, it has led to crystallization of a new paradigm "computing by folding and recombination." Also, the study of gene assembly has led to a discovery that interesting computations take place during important life processes in ciliates, and, moreover, some of these computations use a basic data structure of computer science, viz., linked lists, which are based on the idea of matching the outgoing pointer (address) of a piece of data with the incoming pointer (address) of the "next" piece of data (see, e.g., [35]). Furthermore, this use of linked lists is quite sophisticated, because pointers themselves are part of the information used for gene expression! Most important, the study of gene assembly contributes to a broader and deeper understanding of the notion of computation.

We do believe that research on the computational nature of gene assembly is only in its initial phase. Many important questions must still be asked (and answered!) before this process is well understood.

## 16.2 Gene Assembly Strategies

An important research area is the study of successful gene assembly strategies. In research presented in this book an assembly strategy consists of *sequential steps*, where each sequential step consists of just *one* application of *one* rule (corresponding to one of our three molecular operations); hence we have considered mainly *sequential strategies*. *Parallel strategies* correspond to sequences of steps where a single step may consist of many applications of many rules. Clearly, one is interested in these more general strategies (which also include sequential strategies) as they most probably correspond to the assembly strategies used by ciliates. However, since in this book we have discussed *effects* of assembly strategies (e.g., patterns of intermediate genes), considering sequential strategies was sufficient (and technically simpler) for our purposes. For example, if an intermediate pattern is generated by a parallel assembly strategy (where several applications of several rules may be combined in one step), then this pattern will also be generated by a sequential strategy, albeit more steps will be needed. Since a sequential strategy is a special case of a parallel strategy, this implies that in investigating patterns of intermediate genes one may restrict oneself to sequential strategies only.

In order to understand the gene assembly process one needs to study (the structure of) gene assemblies themselves and not only their effects. For example, the following questions are natural to ask.

- When can different rules form a single step of an assembly strategy?
- What makes a given strategy more feasible or more parsimonious than another one?
- How are different strategies to be compared, e.g., when are two strategies "equivalent"?

Answers to such questions may come from the understanding of the formal constraints as given, e.g., by the overlap graphs as well as from the understanding of biological constraints that perhaps are not expressible by overlap graphs.

## 16.3 Scope of the Operations

Another important research area is concerned with narrowing the scope of our operations. In general, when an operation $\varphi$ is applied to a molecule, say $\alpha$, then the structure of the parts of $\alpha$ that are affected by $\varphi$ can be quite involved. It is therefore desirable to consider such applications of $\varphi$ where the affected structures are "simple." Considering only simple cases in the process of gene assembly makes the analysis of this process perhaps more realistic. In this section we shall discuss simple operations somewhat informally. In particular, we do not consider simple rules that involve markers.

As discussed in Chap. 6, we consider either a simple or a boundary (application of the) $ld$-rule. This restriction is made in order to ensure that the application of the $ld$-rule will not separate the MDSs of a micronuclear or an intermediate gene into two different molecules. If this were the case, some of the MDSs would be lost (in our intramolecular model), and the macronuclear gene could not be assembled. Note that a simple $ld$-rule $\varphi$ (see Fig. 7.1) implies that the part of the molecule affected by the application of $\varphi$ is as simple as it can be: there is only an IES between the two occurrences of the pointer used by $\varphi$.

An application of the $hi$-rule is simple if the part of the molecule that separates two copies of a pointer in an inverted repeat contains only one IES (and almost a whole MDS). More formally, in terms of MDS descriptors, this corresponds in rules (h1) and (h2) from Sect. 7.1 to $\delta_2 = \Lambda$; see Fig. 16.1.

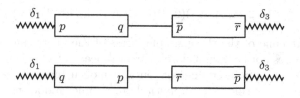

**Fig. 16.1.** The MDS/IES structures where the *simple hi*-rule is applicable. Between the two MDSs there is only an IES.

An application of the *dlad*-rule involving pointers $p$ and $q$ is simple if

- there is no IES between the first copy of $p$ and the first copy of $q$, and
- the part of the molecule between the second copy of $p$ and the second copy of $q$ is an IES.

More formally, in terms of MDS descriptors, the simple cases correspond to Case (d6), Sect. 7.1 with $\delta_4 = \Lambda$ and Case (d7), Sect. 7.1 with $\delta_2 = \Lambda$ (see Fig. 16.2).

Biological investigation including experimentation is needed in order to support the feasibility of simple applications as formulated above. Also, formal

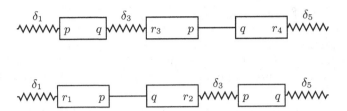

**Fig. 16.2.** The MDS/IES structures where *simple dlad*-rules is applicable. *Straight line* denotes one IES.

investigation of the gene assembly model restricted to simple applications is required. It is our experience, so far, that the mathematical investigation of the simple model is not simple! That the situation for the simple model is quite different from the general model is indicated by the following.

A central property of the three molecular operations *ld*, *hi*, and *dlad* is that they can assemble any macronuclear gene from its micronuclear form. This is expressed, e.g., in terms of MDS descriptors by Theorem 7.1: each realistic MDS descriptor has a successful assembly strategy. However, this is no longer true if one considers only simple applications of these operations, as shown by the following example.

*Example 16.1.* Consider the realistic MDS descriptor

$$\delta = (b, 2)(4, 5)(7, 8)(2, 3)(5, 6)(8, 9)(3, 4)(6, 7)(9, e).$$

We show that none of the three simple forms of the operations is applicable to $\delta$. Since $\delta$ does not contain a simple direct repeat pattern, *ld* is not applicable to $\delta$. Also, for no pointer $p$ does $\delta$ contain an occurrence of the inversion $\bar{p}$, and so the simple *hi*-rule is not applicable to $\delta$. Since $\delta$ does not contain any simple alternating direct repeat, the simple *dlad*-rule is not applicable to $\delta$.

□

## 16.4 Pointer Alignment

A central research problem concerns the mechanism of pointer alignment and MDS splicing. Since pointers are rather short base sequences, they necessarily occur randomly quite often in the micronuclear and intermediate molecules. Therefore any biological mechanism that uses pointers in the process of gene assembly must distinguish between "true" and "false" pointers to provide unambiguous pointer alignments. Moreover, such a mechanism must also reflect the fact that in order for gene assembly to proceed to completion, the joining of MDSs by pointer recombination must be irreversible.

A recently proposed model for the recombination mechanism, based on experimental observations suggested that old macronuclear genes serve as templates in gene assembly. A virtue of this *template-guided recombination model* is that it provides satisfactory solutions to several troubling problems: it identifies the correct copies of the pointer sequences, defines precisely the boundaries between MDSs and adjacent IESs, and provides for irreversible recombination.

Further biological investigation of this template-guided recombination model including experimentation is required (and, indeed, is already proceeding) before the model can be generally accepted as the mechanism for guiding gene assembly. In order to understand this model and accept its validity, formal studies of it are also desirable.

## Notes on References

- Simple forms of the molecular operations *ld*, *hi*, and *dlad* were introduced by Ehrenfeucht et al. [19]. These restricted operations have been shown to be sufficient for assembling all experimentally known cases of micronuclear genes, see Prescott et al. [51].
- The template model referred to in Sect. 16.4 is from Prescott et al. [52].

# References

1. Adleman, L.M., Molecular computation of solutions to combinatorial problems. *Science* **226** (1994), 1021–1024
2. Alberts, B., Bray, D., Lewis, J., Raff, M., Johnson, A., *Essential Cell Biology: An Introduction to the Molecular Biology of the Cell*, Garland Pub (1998)
3. Anderson, E., Chrobak, M., Noga, J., Sgall, J., and Woeginger, G., Solution of a problem in DNA computing. *Theoret. Comput. Sci.* **287** (2002) 387–391
4. Berman, P., and Hannenhalli, S., Fast sorting by reversals. *Combinatorial Pattern Matching, Lecture Notes in Comput. Sci.* **1072** (1996) 168–185
5. Bouchet, A., Circle graphs. *Combinatorica* **7** (1987) 243–254
6. Bouchet, A., Circle graph obstructions. *J. Combin. Theory Ser B* **60** (1994) 107–144
7. Caprara, A., Sorting by reversals is difficult. In S. Istrail, P. Pevzner and M. Waterman (eds.) *Proceedings of the 1st Annual International Conference on Computational Molecular Biology* (1997) pp. 75–83
8. Daley, M., *Computational Modeling of Genetic Processes in Stichotrichous Ciliates*. PhD thesis, University of London, Ontario, Canada (2003)
9. Daley, M., and Kari, L., Some properties of ciliate bio-operations. *Lecture Notes in Comput. Sci.* **2450** (2003) 116–127
10. Daley, M., Ibarra, O. H., and Kari, L., Closure propeties and decision questions of some language classes under ciliate bio-operations. *Theoret. Comput. Sci.*, to appear
11. Dassow, J., Mitrana, V., and Salomaa, A., Operations and languages generating devices suggested by the genome evolution. *Theoret. Comput. Sci.* **270** (2002) 701–738
12. de Fraysseix, H., Local complementation and interlacement graphs. *Discrete Math.* **33** (1981) 29–35
13. de Fraysseix, H., A characterization of circle graphs. *European J. Combin.* **5** (1984) 223–238
14. de Fraysseix, H., and Ossona de Mendez, P., A short proof of a Gauss problem. *Lecture Notes in Comput. Sci.* **1353** (1997) 230–235
15. Ehrenfeucht, A., Harju, T., Petre, I., Prescott, D. M., and Rozenberg, G., Formal systems for gene assembly in ciliates. *Theoret. Comput. Sci.* **292** (2003) 199–219

16. Ehrenfeucht, A., Harju, T., Petre, I., and Rozenberg, G., Patterns of micronuclear genes in cliates. *Lecture Notes in Comput. Sci.* **2340** (2002) 279–289

17. Ehrenfeucht, A., Harju, T., Petre, I., and Rozenberg, G., Characterizing the micronuclear gene patterns in ciliates. *Theory of Comput. Syst.* **35** (2002) 501–519

18. Ehrenfeucht, A., Harju, T., and Rozenberg, G., Gene assembly through cyclic graph decomposition. *Theoret. Comput. Sci.* **281** (2002) 325–349

19. Ehrenfeucht, A., Petre, I., Prescott, D. M., and Rozenberg, G., Universal and simple operations for gene assembly in ciliates. In: V. Mitrana and C. Martin-Vide (eds.) *Words, Sequences, Languages: Where Computer Science, Biology and Linguistics Come Across*, Kluwer Academic, Dortrecht, (2001) pp. 329–342

20. Ehrenfeucht, A., Petre, I., Prescott, D. M., and Rozenberg, G., String and graph reduction systems for gene assembly in ciliates. *Math. Structures Comput. Sci.* **12** (2001) 113–134

21. Ehrenfeucht, A., Petre, I., Prescott, D. M., and Rozenberg, G., Circularity and other invariants of gene assembly in cliates. In: M. Ito, Gh. Păun and S. Yu (eds.) *Words, semigroups, and transductions*, World Scientific, Singapore, (2001) pp. 81–97

22. Ehrenfeucht, A., Prescott, D. M., and Rozenberg, G., Computational aspects of gene (un)scrambling in ciliates. In: L. F. Landweber, E. Winfree (eds.) *Evolution as Computation*, Springer, Berlin, Heidelberg, New York (2001) pp. 216–256

23. Ghosh, A., Tsutsui, S. (eds.), *Advances in Evolutionary Computing: Theory and Applications*, Natural Computing Series, Springer, Berlin Heidelberg New York (2003)

24. Hannenhalli, S., and Pevzner, P. A., Transforming cabbage into turnip (Polynomial algorithm for sorting signed permutations by reversals). In: *Proceedings of the 27th Annual ACM Symposium on Theory of Computing* (1995) pp. 178–189

25. Harju, T., Petre, I., and Rozenberg, G., Formal properties of gene assembly: Equivalence problem for overlap graphs. To appear.

26. Harju, T., and Rozenberg, G., Computational processes in living cells: gene assembly in ciliates. *Lecure Notes in Comput. Sci.* **2450** (2003) 1–20

27. Haykin, S. S., *Neural Networks: A Comprehensive Foundation*, Prentice Hall (1998)

28. Head, T., Formal language theory and DNA: an analysis of the generative capacity of specific recombinant behaviors. *Bull. Math. Biol.* **49**(6) (1987) 737–190

29. Hirvensalo, M., *Quantum Computing*. Natural Computing Series, Springer-Verlag, Berlin Heidelberg New York (2001)

30. Jonoska, N., and Seeman, N. C. (eds.), *DNA Computing*, 7th International Workshop on DNA Based Computers, DNA 7, *Lecture Notes in Comput. Sci.* **2340**, Springer, Berlin Heidelberg New York (2002)

31. Kaplan, H., Shamir, R., and Tarjan, R. E., A faster and simpler algorithm for sorting signed permutations by reversals. *SIAM J. Comput.* **29** (1999) 880–892

32. Kari, J., and Kari, L. Context free recombinations. In: C. Martin-Vide and V. Mitrana (eds.) *Where Mathematics, Computer Science, Linguistics, and Biology Meet*, Kluwer Academic, Dordrecht, (2000) 361–375

33. Kari, L., Kari, J., and Landweber, L. F., Reversible molecular computation in ciliates. In: J. Karhumäki, H. Maurer, G. Păun and G. Rozenberg (eds.) *Jewels are Forever*, Springer, Berlin HeidelbergNew York (1999) pp. 353–363

34. Kari, L., and Landweber, L. F., Computational power of gene rearrangement. In: E. Winfree and D. K. Gifford (eds.) *Proceedings of DNA Bases Computers, V* American Mathematical Society (1999) pp. 207–216

35. Knuth, D.E., *The Art of Computer Programming*, vol. 1, *Fundamental Algorithms*, 3rd edn, Addison-Wesley, Reading, MA (1997)

36. Kotzig, A., Moves without forbidden transitions in graph. *Mat. casopis* **18** (1968) 76–80

37. Landweber, L. F., and Kari, L., The evolution of cellular computing: Nature's solution to a computational problem. In: *Proceedings of the 4th DIMACS Meeting on DNA-Based Computers*, Philadelphia, PA (1998) pp. 3–15

38. Landweber, L. F., and Kari, L., Universal molecular computation in ciliates. In: L. F. Landweber and E. Winfree (eds.) *Evolution as Computation*, Springer, Berlin Heidelberg New York (2002)

39. Lesk, A. M., *Introduction to Bioinformatics*, Oxford University Press, Oxford (2002)

40. Nanninga, N. (ed.), *Molecular Cytology of Escherichia coli*, Academic Press, London (1985)

41. Păun, G., Rozenberg, G., and Salomaa, A., *DNA Computing*, Springer, Berlin Heidelberg New York (1998)

42. Pevzner, P. A., DNA physical mapping and alternating Eulerian cycles in colored graphs. *Algorithmica* **13** (1995) 77–105

43. Pevzner, P. A., *Computational Molecular Biology: An Algorithmic Approach*. MIT Press (2000)

44. Prescott, D. M., *Cells: Principles of Molecular Structure and Function*, Jones and Barlett, Boston (1988)

45. Prescott, D. M., Cutting, splicing, reordering, and elimination of DNA sequences in hypotrichous ciliates. *BioEssays* **14** (1992) 317–324

46. Prescott, D. M., The unusual organization and processing of genomic DNA in hypotrichous ciliates. *Trends in Genet.* **8** (1992) 439–445

47. Prescott, D. M., The DNA of ciliated protozoa. *Microbiol. Rev.* **58**(2) (1994) 233–267

48. Prescott, D. M., Genome gymnastics: unique modes of DNA evolution and processing in ciliates. *Nat. Rev. Genet.* 1(3) (2000) 191–198

49. Prescott, D. M., Bostock, C. J., Murti, K. G., Lauth, M. R., Gamow, E., DNA in ciliated protozoa. I. Electron microscopy and sedimentation analysis of macronuclear and micronuclear DNA of *Stylonychia mytilus*. *Chromosoma* **34** (1971) 355–366

50. Prescott, D. M., and DuBois, M., Internal eliminated segments (IESs) of Oxytrichidae. *J. Eukariot. Microbiol.* **43** (1996) 432–441

51. Prescott, D. M., Ehrenfeucht, A., and Rozenberg, G., Molecular operations for DNA processing in hypotrichous ciliates. *Europ. J. Protistology* **37** (2001) 241–260

52. Prescott, D. M., Ehrenfeucht, A., and Rozenberg, G., Template-guided re-combination for IES elimination and unscrambling of genes in stichotrichous ciliates. Technical Report 2002-01, LIACS, Leiden University (2002)

53. Prescott, D. M., and Rozenberg, G., How ciliates manipulate their own DNA – A splendid example of natural computing. *Natural Computing* **1** (2002) 165–183

54. Prescott, D. M., and Rozenberg, G., Encrypted genes and their reassembly in ciliates. In: M. Amos (ed.) *Cellular Computing*, Oxford University Press, Oxford (2003)

55. Smith, W. D., and Schweitzer, A., DNA computers in vitro and vivo. Technical report, NEC Research Institute (1995)

56. Tucker, A. C., A new applicable proof of the Euler circuit theorem. *Amer. Math. Monthly* **83** (1976) 638–640

57. Weiss, R., Basu, S., Hooshangi, S., Kalmbach, A., Karig, D., Mehreja, R., and Netravali, I., Genetic circuit building blocks for cellular computation, communications and signal processing. *Natural Computing* **2** (2003) 47–84

58. West, D. B., *Introduction to Graph Theory*, Prentice Hall, Upper Saddle River, NJ (1996)

# Index

## Natural Computing Series

W.M. Spears: Evolutionary Algorithms. The Role of Mutation and Recombination. XIV, 222 pages, 55 figs., 23 tables. 2000

H.-G. Beyer: The Theory of Evolution Strategies. XIX, 380 pages, 52 figs., 9 tables. 2001

L. Kallel, B. Naudts, A. Rogers (Eds.): Theoretical Aspects of Evolutionary Computing. X, 497 pages. 2001

M. Hirvensalo: Quantum Computing. 2nd ed., XI, 190 pages. 2004 (first edition published in the series)

G. Păun: Membrane Computing. An Introduction. XI, 429 pages, 37 figs., 5 tables. 2002

A.A. Freitas: Data Mining and Knowledge Discovery with Evolutionary Algorithms. XIV, 264 pages, 74 figs., 10 tables. 2002

H.-P. Schwefel, I. Wegener, K. Weinert (Eds.): Advances in Computational Intelligence. VIII, 325 pages. 2003

A. Ghosh, S. Tsutsui (Eds.): Advances in Evolutionary Computing. XVI, 1006 pages. 2003

L.F. Landweber, E. Winfree (Eds.): Evolution as Computation. DIMACS Workshop, Princeton, January 1999. XV, 332 pages. 2002

M. Amos: Theoretical and Experimental DNA Computation. Approx. 200 pages. 2004

A.E. Eiben, J.E. Smith: Introduction to Evolutionary Computing. XV, 299 pages. 2003

G. Ciobanu (Ed.): Modelling in Molecular Biology. Approx. 300 pages. 2004

R. Paton, H. Bolouri, M. Holcombe, J. H. Parish, R. Tateson (Eds.): Computation in Cells and Tissues. Approx. 350 pages. 2004

L. Sekanina: Evolvable Components. From Theory to Hardware Implementations. XVI, 194 pages. 2004

A. Ehrenfeucht, T. Harju, I. Petre, D.M. Prescott, G. Rozenberg: Computation in Living Cells. XIV, 202 pages. 2004